T0292007

Wireless Communication Networks Supported by Autonomous UAVs and Mobile Ground Robots

Wireless Communication Networks Supported by Autonomous UAVs and Mobile Ground Robots

Hailong Huang
Department of Aeronautical and Aviation Engineering
The Hong Kong Polytechnic University
Kowloon, Hong Kong

Andrey V. Savkin
School of Electrical Engineering and Telecommunications
University of New South Wales
Sydney, NSW, Australia

Chao Huang
Department of Industrial and Systems Engineering
The Hong Kong Polytechnic University
Kowloon, Hong Kong

ACADEMIC PRESS
An imprint of Elsevier

Academic Press is an imprint of Elsevier
125 London Wall, London EC2Y 5AS, United Kingdom
525 B Street, Suite 1650, San Diego, CA 92101, United States
50 Hampshire Street, 5th Floor, Cambridge, MA 02139, United States
The Boulevard, Langford Lane, Kidlington, Oxford OX5 1GB, United Kingdom

Copyright © 2022 Elsevier Inc. All rights reserved.

MATLAB® is a trademark of The MathWorks, Inc. and is used with permission.
The MathWorks does not warrant the accuracy of the text or exercises in this book.
This book's use or discussion of MATLAB® software or related products does not constitute endorsement or
sponsorship by The MathWorks of a particular pedagogical approach or particular use of the MATLAB®
software.

No part of this publication may be reproduced or transmitted in any form or by any means, electronic or
mechanical, including photocopying, recording, or any information storage and retrieval system, without
permission in writing from the publisher. Details on how to seek permission, further information about the
Publisher's permissions policies and our arrangements with organizations such as the Copyright Clearance
Center and the Copyright Licensing Agency, can be found at our website: www.elsevier.com/permissions.

This book and the individual contributions contained in it are protected under copyright by the Publisher
(other than as may be noted herein).

Notices

Knowledge and best practice in this field are constantly changing. As new research and experience broaden
our understanding, changes in research methods, professional practices, or medical treatment may become
necessary.

Practitioners and researchers must always rely on their own experience and knowledge in evaluating and
using any information, methods, compounds, or experiments described herein. In using such information or
methods they should be mindful of their own safety and the safety of others, including parties for whom they
have a professional responsibility.

To the fullest extent of the law, neither the Publisher nor the authors, contributors, or editors, assume any
liability for any injury and/or damage to persons or property as a matter of products liability, negligence or
otherwise, or from any use or operation of any methods, products, instructions, or ideas contained in the
material herein.

Library of Congress Cataloging-in-Publication Data
A catalog record for this book is available from the Library of Congress

British Library Cataloguing-in-Publication Data
A catalogue record for this book is available from the British Library

ISBN: 978-0-323-90182-6

For information on all Academic Press publications
visit our website at https://www.elsevier.com/books-and-journals

Publisher: Mara Conner
Acquisitions Editor: Tim Pitts
Editorial Project Manager: Leticia M. Lima
Production Project Manager: Prem Kumar Kaliamoorthi
Designer: Miles Hitchen

Typeset by VTeX

Contents

Preface

Autonomous vehicles, such as unmanned aerial vehicles (UAVs) and ground mobile robots, have become a modern tool to help humans conduct various missions, which include, but are not limited to, inspection, precise agriculture, search and rescue, and entertainment. In recent years, UAVs and ground mobile robots have been widely used to assist wireless communication systems. In particular, they can play the role of either data sinks to collect sensory data from sensor nodes deployed in a field or access points to extend the coverage communication service. This book is primarily a research monograph that presents, in a detailed and unified manner, the recent advancements relevant to the application of autonomous vehicles in wireless communication systems. The main intended audience for this monograph is postgraduate and graduate students, as well as professional researchers and industry practitioners that are working in a variety of areas such as robotics, control engineering, and computer science. This book is essentially self-contained, and a prerequisite is a familiarity with basic undergraduate-level mathematics. The approaches presented are discussed to a great extent and illustrated by examples. We hope that readers find this monograph interesting and useful and gain a deeper insight into the challenging issues in the field. Moreover, in the book, we have made comments on some open issues, and we encourage readers to explore them further. The material in this book is the result of the authors' research between 2016 and 2021. Some of its parts have separately appeared in journal and conference papers. The manuscript integrates them into a unified whole, highlights the connections between them, supplements them with new original findings of the authors, and systematically presents the entire material.

In preparation for this manuscript, we would like to acknowledge the financial support we received from the Australian Research Council. We also received funding from the Australian Government, via grant AUSMURIB000001 associated with ONR MURI grant N00014-19-1-2571.

Hailong Huang
Andrey V. Savkin
Chao Huang

Chapter 1

Introduction

1.1 Autonomous vehicles in wireless communication networks

Autonomous vehicles, such as mobile ground robots and unmanned aerial vehicles (UAVs), have reshaped our modern life. Thanks to lightweight and low-cost components, autonomous vehicles have become new means to conduct dangerous and time-consuming missions. Typical examples include the inspection of disaster areas [1] and power line inspection [2]. In recent years, mobile ground robots and UAVs have been widely used to assist wireless communications. Specifically, they can play the role of either data sinks to collect sensory data from sensor nodes deployed in a field [3] or the access points to extend the coverage communication service [4]. In such applications, the design of approaches generally follows a four-layer framework, which consists of scheduling, path planning, motion control, and communication protocols.

Scheduling. Given a number of autonomous vehicles to execute the expected mission, the scheduling of vehicles determines the overall performance. The scheduling problem is one of the best-known combinatorial optimization problems [5] and has attracted lots of interest. The scheduling of vehicles refers to the task allocation to each vehicle, so that a certain metric, e.g., the cost or the completion time, is optimized. Under different scenarios, in the scheduling of vehicles some constraints also need to be considered. For example, one task can be conducted only if another is completed first, and one task must be conducted within a certain time window.

Path planning. Once a vehicle is assigned a task, the next issue to address is the path planning problem. In the context of autonomous vehicles, a mobile ground robot or a UAV is usually assigned to visit a set of positions. A commonly used approach to address the path planning problem is to formulate the problem as a traveling salesman problem (TSP) [6]. Typically, TSP aims at finding the shortest path for the salesman to visit the given sites exactly once. Here, the shortest path may refer to the path with the shortest completing time, the least energy consumption, etc. Since autonomous vehicles may have some limited mobility, such as not being able to make sharp turns [7], such mobility constraints need to be taken into account in the path planning process. In particular, this approach is based on the knowledge of the environment [8], and the corresponding planner is often called the global planner. Generally, global planners guarantee not only collision avoidance but also achieve a global navigation

Wireless Communication Networks Supported by Autonomous UAVs and Mobile Ground Robots
https://doi.org/10.1016/B978-0-32-390182-6.00006-9
Copyright © 2022 Elsevier Inc. All rights reserved.

objective if certain general assumptions about the environment are met. Different from global planners, local planners only use the onboard sensors to detect a part of the environment and plan a short-horizon path iteratively [9].

Motion control. With a planned path, the next issue is how to steer the vehicle so that it precisely follows the path, which is a typical reference tracking problem. The reference tracking problems are often described as optimization problems with certain constraints. Such problems are often with the objective of minimizing the error between the reference and the actual system output (i.e., trajectory) and the magnitude of the control inputs over a given horizon. Minimizing the error indicates the goal of precise tracking, while minimizing the magnitude of control inputs indicates the goal of using the minimum effort to achieve the reference tracking. Regarding the constraints, the commonly considered one is the mobility constraint of the vehicles. In practice, an autonomous vehicle only accepts control inputs within certain ranges, and such a vehicle is called underactuated. Another constraint comes from obstacles. For safety purposes, an autonomous vehicle must keep away from any obstacle by a certain distance [10]. Researchers have developed different methods for motion control, and typical examples include model predictive control (MPC) [11] and sliding mode control (SMC) [12].

Communication protocol. When a vehicle reaches a certain position or when it is on the way, some communication protocols need to be in place for the communications between the users and the vehicle. The simplest case is that only a single user needs to communicate with the vehicle at any time. Then, the vehicle can send a calling message, and the user uploads its data to the vehicle. If necessary, the vehicle may return some required data to the user. The complex situation is that a vehicle needs to serve multiple users. The scheduling process determines which users a vehicle should serve. A common strategy is the popular time-division multiple access (TDMA) [13], which is a channel access method for shared-medium networks. TDMA allows multiple users to share a certain channel by dividing the signal into different time slots. The users transmit in rapid succession using their own time slots. Another well-known strategy is called the frequency-division multiple access (FDMA) [14]. FDMA allows multiple users to send data through a single communication channel by dividing the bandwidth of the channel into separate nonoverlapping frequency subchannels and allocating each subchannel to one user. Moreover, for some large networks, not every user may be allowed to directly communicate with the vehicles. In this situation, multihop communications should be designed either deterministically [15] or opportunistically [16].

Within the aforementioned framework, to achieve good performance, the following key issues need to be considered carefully. Firstly, many autonomous vehicles are constrained in energy capacity, which results in a limited operation time. For example, most low-cost multirotor UAVs can only fly for about 30 minutes with a fully charged battery. The direct solution is to develop a high-capacity battery [17]. The second approach is to develop energy-efficient planning and control methods so that the vehicles can operate relatively long under the constraints. Another approach is to install solar panels on the vehicles so that they can harvest solar energy during the operation [18]. In the last approach, trajectory planning is vital especially in areas with constructions such as urban areas or mountainous regions, because the constructions and mountains may create shadows preventing the harvest of solar energy. Secondly, the mobility of the users needs to be well considered. In the case where there is a limited number of vehicles to serve a number of moving users, the users can only be served periodically. The management of autonomous vehicles is a challenging task, especially when the movements of the users are unknown in advance. Thirdly, when multiple vehicles cooperate, they need to share the operating space. This is likely to lead to conflict in terms of task assignment. For example, a user which is assigned to a vehicle for service may enter the service range of another vehicle, since the user can move. In this case, if the task assignment is not well scheduled, the former vehicle may waste its resource while the user is served by the latter vehicle. Moreover, when the vehicles share the operating space, they may also have a high probability of collision, especially for the usage of UAVs [19]. To achieve the expected performance, these issues should be carefully considered in the design of approaches, and reactive approaches are preferred to deal with dynamic situations.

This monograph aims at overcoming some of the aforementioned deficiencies in the previous research. For the usage of mobile ground robots, this book not only discusses the scenario where the robots can move freely in the field but also investigates the case where the robots move only on some fixed routes. For the former, we focus on the path planning of the robots so that they can efficiently collect sensory data from sensor nodes. For the latter, we focus on the routing protocol design especially for the scenarios where urgent data should be collected within a given threshold. To address the energy limitation issue about UAVs, we pay attention to the solution of installing solar panels on UAVs. We present a multiple-objective optimization framework, where several important objective functions can be considered at the same time. These objective functions include but are not limited to the amount of harvested solar energy, the total time for secure communication, and the length of the UAV's trajectory. Moreover, this book covers the latest approaches to reactive navigation. We consider the case where multiple UAVs serve multiple targets, and we focus on the coordination algorithms.

This book is problem-oriented, not technique-oriented. So, each chapter is self-contained and is devoted to a detailed discussion of an interesting problem that arises in the rapidly developing area of mobile robot and UAV-assisted wireless communication networks. We present the relevant approaches from the viewpoint of control systems. Thus, in Chapters 4 to 10, we first present the system models and then formulate the problems of interest, which is followed by the proposed approaches to address the problems. Finally, we present computer simulation results to illustrate the effectiveness of the approaches.

1.2 Overview and organization of the book

In Chapters 2 and 3, we discuss the development of the usage of mobile robots and UAVs in wireless communication networks. Specifically, in Chapter 2, we present a review of techniques related to wireless communication networks supported by mobile ground robots. We highlight some directions in which available approaches may be improved, and we cover typical approaches on the deployment, navigation, and control of mobile ground robots to support the operation of the wireless sensor networks (WSNs). In Chapter 3, we focus on the usage of UAVs in wireless communication networks.

In Chapter 4, we focus on the path planning problem for mobile ground robots to support the operation of the WSNs. We focus on designing paths for mobile robots. Considering the mobility constraints of mobile robots, we introduce the concept of viable path, which combines the concerns of both robots and sensor networks. We formulate the problem of planning the shortest viable path for a single robot as a variant of the Dubins TSP with neighborhoods (DTSPN). Accordingly, we develop a shortest viable path planning (SVPP) algorithm. We further consider the problem of planning viable paths for multiple robots and present a k-shortest viable path planning (k-SVPP) algorithm. As the constraints of mobile robots and sensor networks are both taken into account in the path planning phase, the created paths enable the robots to effectively and efficiently collect data from sensor nodes.

In Chapter 4, we focus on a scenario where mobile robots can move freely in the field. However, it may meet difficulty when applied to real applications. The main reason is that in real applications, the environment in which the mobile robots are moving is quite complex. It has not only obstacles, as mentioned in Chapter 4, but also other types of restrictions, for example, the mobile robots have to move only on roads in urban areas. Therefore, it is necessary to study another mobility pattern, i.e., constrained mobility. Thus, in Chapter 5, we focus on how this kind of mobility can assist data collection in WSNs.

Chapters 6–10 focus on the usage of UAVs in wireless communication. Chapter 6 discusses how to plan the trajectory of a solar-powered UAV under a cloudy condition to secure the communication between the UAV and a target ground node against multiple eavesdroppers. We propose a new 3D UAV trajectory optimization model by taking into account the UAV energy consumption, solar power harvesting, eavesdropping, and no-fly zone avoidance. A rapidly exploring random tree (RRT) method is developed to construct the UAV trajectory.

Chapter 7 considers a multiobjective path planning problem, which jointly considers the maximization of the residual energy of the solar-powered UAV at the end of the mission, the maximization of the time period in which the UAV can securely communicate with the intended node, and the minimization of the time to reach the destination. We pay attention to the impact of the buildings in urban environments, which may block the transmitted signals and also create some shadow region where the UAV cannot harvest energy. An RRT-based path planning scheme is presented. This scheme captures the nonlinear UAV motion model and is computationally efficient considering the randomness nature. From the generated tree, a set of possible paths can be found. We evaluate the security of the wireless communication, compute the overall energy consumption as well as the harvested amount for each path, and calculate the time to complete the flight. Compared to a general RRT scheme, the proposed method enables a large time window for the UAV to securely transmit data.

In Chapters 8 to 10, we turn to approaches to deploying and navigating multiple UAVs for wireless communication, rather than a single UAV, as in Chapters 6 and 7. In Chapter 8, each UAV carries a battery with limited initial energy and can provide connectivity with a ground user that is within a certain range. When the energy consumption of a UAV depends on its altitude, minimizing the energy consumption and maximizing the number of serviced users are two contradictory goals because to have a larger coverage, a UAV needs to fly higher, which leads to more energy consumption. Thus, there should be a balance between them. A constrained optimization problem taking these two objectives into account is formulated subject to some energy and connectivity constraints. A control system containing a movement decision maker (MDM) is designed. A decentralized navigation algorithm implemented on UAV is proposed. The algorithm navigates each UAV to a new position in 3D space that contributes more to the coverage.

In Chapter 9, we consider the coverage of ground users in urban areas. The main difference lies in the line of sight (LoS), which has been discussed in Chapter 7 for a single UAV scenario. In urban areas, the wireless coverage highly depends on whether a UAV base station (UAV-BS) has LoS with a user. Although a user is within the coverage radius of a UAV-BS, buildings may block data transmission, which significantly reduces the transmission quality. In this chapter, we focus on the 3D deployment problem of UAV-BSs to serve ground users in a given area. We formulate an optimization problem to find the optimal 3D positions for UAV-BSs with the objective of maximizing the number of covered users, subject to the constraints that UAV-BSs should be deployed at safe positions and the covered users receive acceptable quality of service. We analyze the difficulty of such a problem and show that it is NP-hard. A greedy algorithm is developed with computational complexity analysis. Extensive computer simulations are presented to illustrate the effectiveness of the proposed algorithm and a comparison with a baseline algorithm is provided to assess the performance gains.

In the last chapter, we consider using UAVs to provide wireless communication services to users in urban areas, and we focus on the deployment problem. In this chapter, we consider using solar-powered UAVs to serve a group of moving users. In particular, we consider the scenario where the number of users is larger than that of UAVs, and the users spread in the environment so that the UAVs need to carry out periodical surveillance. The existence of tall buildings in urban environments brings new challenges to the periodical surveillance mission. They may not only block the LoS between a UAV and a user but also create some shadow region, so that the communication quality may become unsatisfactory, and the UAV may not be able to harvest energy from the sun. The periodical service problem is formulated as an optimization problem to minimize the user revisit time while taking the impact of the urban environment into account. A nearest neighbor-based navigation method is proposed to guide the movements of the UAVs. Moreover, we adopt a partitioning scheme to group users for the purpose of narrowing UAVs' moving space, which further reduces the user revisit time.

References

[1] R.R. Murphy, J. Kravitz, S.L. Stover, R. Shoureshi, Mobile robots in mine rescue and recovery, IEEE Robotics & Automation Magazine 16 (2) (2009) 91–103.

[2] D. Ma, Y. Li, X. Hu, H. Zhang, X. Xie, An optimal three-dimensional drone layout method for maximum signal coverage and minimum interference in complex pipeline networks, IEEE Transactions on Cybernetics (2021) 1–11.

[3] H. Huang, A.V. Savkin, Viable path planning for data collection robots in a sensing field with obstacles, Computer Communications 111 (2017) 84–96.

[4] H. Huang, A.V. Savkin, A method for optimized deployment of unmanned aerial vehicles for maximum coverage and minimum interference in cellular networks, IEEE Transactions on Industrial Informatics 15 (5) (2019) 2638–2647.

[5] R.L. Graham, Bounds for certain multiprocessing anomalies, Bell System Technical Journal 45 (9) (1966) 1563–1581.

[6] J.K. Lenstra, A.R. Kan, Some simple applications of the travelling salesman problem, Journal of the Operational Research Society 26 (4) (1975) 717–733.

[7] C. Wang, A.V. Savkin, M. Garratt, A strategy for safe 3D navigation of non-holonomic robots among moving obstacles, Robotica 36 (2) (2018) 275.

[8] N.A. Vlassis, N.M. Sgouros, G. Efthivoulidis, G. Papakonstantinou, P. Tsanakas, Global path planning for autonomous qualitative navigation, in: Proceedings Eighth IEEE International Conference on Tools with Artificial Intelligence, IEEE, 1996, pp. 354–359.

[9] L. Lapierre, R. Zapata, P. Lepinay, Combined path-following and obstacle avoidance control of a wheeled robot, The International Journal of Robotics Research 26 (4) (2007) 361–375.

[10] M. Rubagotti, M.L. Della Vedova, A. Ferrara, Time-optimal sliding-mode control of a mobile robot in a dynamic environment, IET Control Theory & Applications 5 (16) (2011) 1916–1924.

[11] C. Huang, F. Naghdy, H. Du, Delta operator-based fault estimation and fault-tolerant model predictive control for steer-by-wire systems, IEEE Transactions on Control Systems Technology 26 (5) (2018) 1810–1817.

[12] J. Guldner, V.I. Utkin, Sliding mode control for an obstacle avoidance strategy based on an harmonic potential field, in: Proceedings of 32nd IEEE Conference on Decision and Control, IEEE, 1993, pp. 424–429.

[13] D.D. Falconer, F. Adachi, B. Gudmundson, Time division multiple access methods for wireless personal communications, IEEE Communications Magazine 33 (1) (1995) 50–57.

[14] M. Morelli, C.-C.J. Kuo, M.-O. Pun, Synchronization techniques for orthogonal frequency division multiple access (OFDMA): a tutorial review, Proceedings of the IEEE 95 (7) (2007) 1394–1427.

[15] R. Draves, J. Padhye, B. Zill, Routing in multi-radio, multi-hop wireless mesh networks, in: Proceedings of the 10th Annual International Conference on Mobile Computing and Networking, 2004, pp. 114–128.

[16] S. Biswas, R. Morris, ExOR: opportunistic multi-hop routing for wireless networks, in: Conference on Applications, Technologies, Architectures, and Protocols for Computer Communications, 2005, pp. 133–144.

[17] Z. Pan, L. An, C. Wen, Recent advances in fuel cells based propulsion systems for unmanned aerial vehicles, Applied Energy 240 (2019) 473–485.

[18] J.-S. Lee, K.-H. Yu, Optimal path planning of solar-powered UAV using gravitational potential energy, IEEE Transactions on Aerospace and Electronic Systems 53 (3) (2017) 1442–1451.

[19] Z. Yan, N. Jouandeau, A.A. Cherif, A survey and analysis of multi-robot coordination, International Journal of Advanced Robotic Systems 10 (12) (2013) 399.

Chapter 2

Survey of approaches for wireless communication networks supported by ground robots☆

2.1 Introduction

WSNs, typical wireless communication networks, were primarily motivated by military applications, and have been widely used in civilian applications such as environmental monitoring and intrusion detection. Benefiting from technological developments, more and more devices can be connected, which leads to the concept of Internet of Things (IoT) [1].

A conventional WSN is composed of a number of wireless sensor nodes to sense their surrounding environment and a data sink (also known as base station) [2]. Sensor nodes collect diverse information sources from the physical world and send it to the sink for analysis. Appropriate actions may be executed to respond to the sensed information. In conventional WSNs, both sensor nodes and sinks are static, and the sensory data at a node are transmitted to the sinks via the relay of other nodes. This feature often leads to the energy hole issue. The sensor nodes close to the sinks need to not only transmit their own sensory data but also relay data for other nodes that are far from the sinks. Thus, the former nodes may run out of battery quickly. Consequently, the latter nodes may be isolated from the sinks. This bottleneck prevents the practical application of WSNs. Much research has been conducted to prolong the lifetime of WSNs, and the first part of this chapter reviews the typical approaches.

Since the aforementioned topics continue to be active areas of research, in this chapter, we comprehensively survey the closely relevant work. We pay special attention to the existing approaches on the deployment, navigation, and control of mobile ground robots to support WSNs. The main objective of this chapter is to provide the state of the art, which makes the following chapters easy to understand. The main material of this chapter was originally published in [3].

☆ The main material of this chapter was originally published in Hailong Huang, Andrey V. Savkin, Ming Ding, Chao Huang, Mobile robots in wireless sensor networks: a survey on tasks, Computer Networks 148 (2018) 1–19. Permission from Elsevier for reuse was obtained.

https://doi.org/10.1016/B978-0-32-390182-6.00007-0
Copyright © 2022 Elsevier Inc. All rights reserved.

2.2 WSNs supported by mobile robots

In this section, we present a literature review of approaches to WSNs supported by mobile robots. The review follows a taxonomy based on the roles of the mobile robots: collection, delivery, and combination [3]. Specifically, the role of collection means the robots are assigned the tasks of collecting data from sensor nodes; the role of delivery means that the robots need to send something to sensor nodes; and the combination role means the robots conduct the aforementioned two operations at the same time.

2.2.1 Collection

The main task of the mobile robots in this category is to collect data from sensor nodes. In this case, a network generally consists of three components: (1) a set of static sensor nodes, (2) one or multiple mobile robots, and (3) one or multiple static sinks. The mobile robots can have different types of mobility, including random mobility (see Section 2.2.1.1), partially controlled mobility (see Section 2.2.1.2), and fully controlled mobility (see Section 2.2.1.3). A static sensor node measures the environment information and can transmit the data to a robot via either single-hop (when it is in proximity) [4–6] or multihop communication to the robot's current location [7–9]. After receiving the sensory data, a robot can either return to a sink to upload the data or immediately send it to a sink via long-distance transmission [4,10] (Fig. 2.1). The reviewed publications are categorized into three groups based on the mobility type of robots: random, partially controllable, and fully controllable. In each group, we present a comprehensive discussion on the representative approaches and summarize their merits and shortcomings.

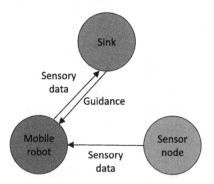

FIGURE 2.1 System structure for data collection.

2.2.1.1 Random mobility

Random mobility is an early mobility type studied in the area of data collection by mobile robots in WSNs. A conventional system design is that the robot–

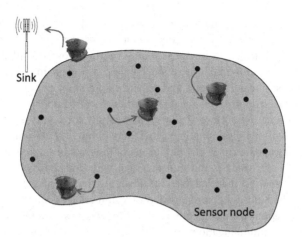

FIGURE 2.2 The system considered in [4].

node communication is activated when the robot comes into the proximity of the sensor node [4]. However, due to the long data collection delay and large storage requirement at sensor nodes, researchers proposed different methods allowing sensor nodes to transmit data through bounded hops [11], in order to reduce the movement of robots. Moreover, multihop robot–node communication strategies have been proposed [7–9,12].

The paper [4] presents one of the first efforts on this topic. A reference architecture with three layers is proposed (Fig. 2.2). Sensor nodes are the first layer. The middle layer contains mobile robots, which travel around the field randomly and pick up data from sensors when in proximity. The top layer is a data sink. Energy efficiency is the major merit of this scheme. Significant power can be saved, because sensor nodes only upload data to mobile robots in single-hop communication. The energy efficiency can be significantly improved in the cases with large data packets, such as videos.

The paper [11] introduces a density-based proactive data dissemination protocol (DEEP), which is targeted at reducing both communication and storage overhead by exploring probabilistic flooding and probabilistic storing. Specifically, the data packets from source nodes propagate by probabilistic flooding until reaching a predefined threshold, i.e., the number of retransmissions of each packet in a given neighborhood. Once an intermediate node receives the packets, it employs a probabilistic storing mechanism to decide if the packet should be stored locally. As a consequence, a mobile robot can collect a representative view of the sensory data by visiting a small subset of sensor nodes, instead of visiting every source node.

The long data collection delay is the main defect of [4,11]. One possible solution is to embed sensor nodes with routing function like conventional WSNs, and at the same time consider the mobile feature. With this idea, a number of mobility-aware routing protocols have been proposed. The key question here is how the source nodes transmit sensory data to robots.

The paper [9] designs a broadcast-based routing protocol called SinkTrail for efficient data gathering in a round manner. Since no mobility pattern is assumed in this scheme, location updates are essential to ensure data transmission. A trail-based approach is presented for the location updates. In particular, the robot stops at some place (called a trail point) for a very short time and broadcasts a trail message containing its ID and a time tag. Whenever a sensor receives this, it updates its route and distance (in terms of hop count) from the sink, and then rebroadcasts the trail message with updated distance information. The main merit is that SinkTrail is GPS-free. The logical network structure gradually formed during the propagation of trail messages can replace accurate geographic locations of sensors obtained by GPS devices. The major concern of SinkTrail is the communication overheads, since it is a broadcast-based protocol. Even by carefully tuning network parameters, the energy consumption on the frequent broadcast of the trail messages could be huge and unbearable. Also, in a large-scale network, sensors that are far away from sinks may suffer from delay induced by broadcasting. With overdue trail messages, data from these sensors may circle around the network and cause significant energy consumption. Or even worse, they may fail to find the sink within the current round, leading to data loss.

The reference [7] proposes a two-tier data dissemination (TTDD) protocol. It is based on flooding and employs a grid structure to reduce the communication overheads caused by the location update. When an event occurs, the sensor nodes nearby collect the sensory data. Then, one of these nodes becomes the source node. The source node constructs a grid overlay with a predefined size and then selects multiple target nodes that are within a certain distance to the grid point for each cell. TTDD divides the sensing area into grids, and thus queries can be dealt with in terms of grids rather than sensors. This scheme reduces more communication overheads than a pure flooding-based approach (e.g., [9]). However, its performance degrades with the increment of network size. The reason is that the communication overheads for maintaining the grid overlay increase dramatically.

The reference [8] proposes a hierarchical cluster-based data dissemination (HCDD) protocol, which divides the network into clusters. Cluster heads (CHs) form a higher-level structure and record the global information such as the mobile robots' IDs and routes towards them. When a mobile robot is inside a cluster, the closest CH propagates a notification to other CHs. Then, the route for source nodes to transmit sensory data is from the source node to its CH and them from its CH to other CHs. This scheme saves the communication overheads caused by constructing and maintaining an infrastructure for each source node (e.g., TTDD). Also, HCDD is a distributed protocol which does not require sensors' location information. Both control overheads and routing overheads have been pushed from the whole network to some specific nodes, i.e., CHs.

The paper [12] explores a rendezvous-based architecture and presents a line-based data dissemination (LBDD) protocol. LBDD combines the rendezvous-based architecture and multihop routing to improve the data transmission efficiency. A virtual region with a certain width is constructed. Sensors within the region function as subsinks, while other sensors operate normally. Subsinks are divided into groups with a certain size. LBDD runs in three phases. Data from ordinary nodes will firstly be transmitted towards the region and buffered by the nearest subsink. Then, the query message from a robot will be forwarded towards the region until it reaches a subsink. This subsink will be responsible for broadcasting the query message in both up and down directions until interesting sensory data are found. Finally, the subsink transmits the data to the robot.

To sum up, the approaches exploring random mobility for the collection task are simple in terms of implementation but suffer from the long delivery delay [4,11]. Two methods to address such issue are embedding sensor nodes with routing function [7–9,12] and enabling sensor nodes to learn the mobility patterns and make predictions for the movements [13,14]. In particular, the approaches exploiting routing protocols are classified into two classes: flat architecture [9] and structure-based [7,8,12]. In terms of scalability, the latter outperforms the former.

2.2.1.2   Partially controllable robots

Partially controllable robots are also widely used in applications. Related publications can be categorized into two groups: stop location-constrained mobility and trajectory-constrained mobility. In the stop location-constrained group, path planning is the first issue that practitioners have to face. However, compared to the case in Section 2.2.1.3, the tasks of path planning is not that complex. Due to the stop location constraints, the number of candidate trajectory is reduced significantly.

Stop location-constrained. In some applications, the mobile robots cannot stop anywhere in the field due to physical constraints. Instead, they can only stop at some predefined locations. A typical example is battle field surveillance, where the mobile robots can only operate in a few safe locations. Another example is habitat monitoring, where only limited places are accessible to robots. The relevant publications generally follow two approaches: real-time uploading [15–18] and delay-tolerant uploading [19,20].

The paper [15] considers the problem of network lifetime maximization, which is achieved by jointly solving two problems: the scheduling problem tho determine the robot stop times and the routing problem to find the appropriate energy-efficient paths. The reference [16] also considers to prolong the network lifetime. The authors propose an integer linear programming (ILP) model to determine the locations of robots from the given candidate locations and the routes from sensor nodes to multiple robots. The main defect of [15] is that there is no relationship of the locations a certain robot should visit during several continuous rounds. Thus, one robot may traverse the network from one side of the sensing field to the other, which may not be practical sometimes. Therefore, it is necessary to enforce constraints on the robots to restrict their movements.

The paper [17] considers the similar problem as [15], and a mixed ILP model is presented. As the solution provided by this model is still a centralized one, the authors further introduce a distributed and localized scheme, i.e., greedy maximum residual energy (GMRE). GMRE only requires the residual energy information of the nodes nearby the robot and it pushes the robot to the energy-rich area. Moreover, the paper [18] extends the MILP model to the multiple-robot case and presents a centralized heuristic algorithm.

The reference [21] presents theoretical results regarding the optimal robot movement. The main result hinges upon a transformation of the joint robot movement and data routing problem to maximize network lifetime from the time domain to the space domain. Based on such transformation, the authors first show that in the case of having predefined stop locations, the optimal stop time at each of the given stop locations can be found.

The aforementioned approaches all consider the model of gathering sensory data in real-time. The paper [19] considers a scenario where delay is allowed. Then, each sensor node does not send the data immediately when available. Instead, it can store the data temporarily and transmit it when the robot is at the position most favorable for achieving the longest lifetime. One drawback of [19] is that the travel time between different locations is ignored. Although this results in a simple problem formulation to obtain a precise numerical solution, the solution is hard to implement in practice. Beyond [19], the paper [20] considers slow mobility and presents a mixed integer nonlinear program (MINLP) to maximize the network lifetime subject to a given allowed delay constraint. Via the analysis of several subproblems of the origin, a polynomial-time approach is proposed. Compared to [19], the results in [20] are more practical due to the consideration of slow mobility.

Trajectory-constrained. In some applications (e.g., in urban areas), the robots' trajectories may also be restricted. For instance, when a robot is carried by a car or a bus, the robot's trajectory should follow the road networks. Such a practical constraint avoids the complex problem of path planning. Then, the research focus is on the cooperation between sensor nodes and robots, i.e., how to efficiently route data packets to robots, given the trajectories or even the arrival time of robots at certain positions.

The paper [22] considers the scenario where a robot is installed on a bus moving on its fixed path periodically and collects data from a set of sensor nodes deployed near the path. It proposes a queuing formulation to model the process of data collection. Moreover, it proposes a communication protocol to assist gathering data by the robot. In a similar context, the paper [23] focuses on the scheduling problem in node–robot transmission and a trade-off between the probability of successful information retrieval and node energy consumption is studied. Different from [23], which considers a sparse network, the reference [24] focuses on dense networks. The authors consider the maximization of data collection throughput by dividing the traversing time into a set of time slots with equal duration and studying the problem of assigning nodes to the time slots. One shortcoming of [22–24] is that they all use single-hop communication, which requires that the sensor nodes are deployed within the communication range of the robot when it moves on the trajectory.

The paper [25] removes the node location assumption, and multihop communication is used. During the movement, the mobile robot broadcasts a message continuously within its communication range. The nodes that can hear this message are called subsinks. Subsinks then forward the message within their communication ranges. When the robot finishes its trip, every node knows its shortest hop distance to a subsink as well as the shortest path towards the subsink. From the robot's next trip, nodes transmit their sensory data to the corresponding subsinks, and when the robot comes, the subsinks upload the data. Compared to [22–24], multihop communication improves the applicability and scalability of system. However, the shortest path-based routing leads to an unbalanced assignment of nodes to subsinks, which may result in that some subsinks having a long contacting time with the robot are associated with a small number of nodes. Thus, the subsinks may not manage to upload all their buffered data. The paper [26] considers the defect of [25], and the authors formulate a constrained assignment problem such that each subsink is associated with an appropriate number of nodes, which enables more data to be collected. The proposed mechanism can improve the overall throughput. But, some nodes may be associated to a far away subsink, thus data relay consumes more energy than in the case of shortest path routing.

Both [25] and [26] are based on a flat network structure, i.e., most sensor nodes participate in data relay, except those without child nodes. The paper [27] considers the same context but clustering is used. The proposed data collection protocol, MobiCluster, is based on a clustering algorithm called unequal routing clustering (URC) [28]. The authors of [28] point out that for a network with a static sink, the clusters near the sink should be smaller than those which are far away from the sink. The reason is as follows. The CHs near the sink have a heavier burden of relaying data compared to the CHs far away. CHs also need to collect data from their cluster members (CMs) within the cluster and aggregate the collected data. Thus, if the cluster size is identical across the network, the CHs near the sink may run out of energy much more quickly than those far away. To avoid this issue, one approach is to construct unequal clusters, i.e., the clusters near the sink have smaller sizes while the clusters far away have larger sizes. The unequal clusters reduce the energy consumption for intracluster communication of CHs near the sink, and these CHs can spend more energy on intercluster communication, i.e., relaying data. One drawback of URC is that it assumes the robot is able to broadcast a packet to all the nodes. Similar to the defect of [22–24], this assumption limits the scalability of URC [28], as MobiCluster [27]. Another drawback is that the cluster size depends on the distance information derived from signal strength, which may not be reliable in

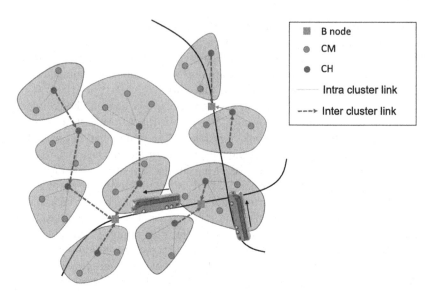

FIGURE 2.3 The system considered in [30], where the so-called B nodes are located at bus stops.

harsh environments. The paper [29] takes into account the nonuniform deployment of sensor nodes, going beyond [27], where only uniform deployment is considered. The authors of [29] follow the idea of URC [27,28] and propose an energy-efficient protocol for WSNs with nonuniformly deployed sensor nodes.

Stop location- and trajectory-constrained. The schemes discussed so far in Section 2.2.1.2 consider one constraint of stop location or trajectory. In practice, there are applications where both of them apply. For example, in [30] a scenario is considered where the mobile sinks are attached to buses. The authors focus on the problem of gathering urgent messages within an allowed latency. They consider using the bus network to collect such urgent message from sensor nodes deployed in urban areas (Fig. 2.3). Since the timetable of buses is available and the stop duration at each bus stop can also be learned through long-term observation, the data collection system by the authors works as follows. Upon detecting an event, a sensor node sends it to its CH. The CH is connected to some special nodes deployed at the bus stops. The CH consults the bus stop nodes for the arrival time of buses as well as their stop duration. Based on this, the CH sends the urgent message to a selected set of bus stop nodes which guarantees that the successful delivery rate is no less than a given threshold. In the meantime, the energy consumption for such process is minimized.

To sum up, for partially controllable mobility, three cases have been considered: stop location-constrained, trajectory-constrained, and stop location- and trajectory-constrained. When the stop location-constrained case is considered, path planning of robots is the most important issue, while when the trajectory-constrained is considered, the collaboration between robots and sensor nodes becomes more significant.

2.2.1.3   Fully controllable robots

Fully controllable robots can either serve as mobile sinks like those in Sections 2.2.1.1 and 2.2.1.2 or as relays. Serving as relays requires less mobility than as mobile sinks. In this subsection, we survey these two aspects.

When the fully controllable robots function as mobile sinks to collect data, practitioners usually need to consider two major issues:

- robot control, including path planning, sojourn time allocation, speed control, etc.,
- data routing, i.e., how to route sensory data to robots.

Regarding the control of robots, the trajectory plan is the issue to be solved first. For trajectory planning, the TSP [31] is often used. In the TSP, a sensor node is regarded as a costumer and the robot is regarded as the salesman. The goal is to find a minimum cost tour, starting and ending at a static base station, that visits each costumer once. No matter what kind of stop time and routing strategies are in use, the data collection delay is decidedly increased due to the physical movement. A variant of the TSP, the TSP with neighborhoods (TSPN) [32], can shorten the tour to some extent by taking into account the communication range of sensor nodes and the robot [33–35]. Moreover, when multiple robots are available, the path planning of multiple robots can be formulated as the k-TSPN [10] or as the vehicle routing problem (VRP) [36].

The paper [35] formulates the trajectory planning problem as a special case of the TSPN, where the neighborhoods are possibly intersecting continuous disks. The authors adopt the multitransmission rate model between a sensor node and a robot, which is a function of the distance between them. A combine-skip-substitute scheme is proposed to shorten the tour length of the robot progressively. The paper [10] considers the scenario where multiple robots are used. The authors formulate the problems as a k-TSPN for the case where the robot uploads data to the sink when they meet and as a k-rooted path cover problem with neighborhoods (k-PCPN) for the case where the robot can upload data to the sink remotely. The k-TSPN aims at finding the k closed tours for the k robots such that the length of the longest tour is minimized, while the k-PCPN aims at finding the k paths for the k robots with the same objective. Heuristic algorithms are proposed to address the problems.

The TSP-based approaches mostly focus on the high level of path planning, while ignoring the low level of control of robots. One significant aspect is that the real mobile platform which can carry mobile sinks usually has a certain constraint. For example, many robots based on the Dubins car model [37] have limited ability to make sharp turns. The TSP for Dubins car (DTSP) was introduced by [38] to find the minimum-length path satisfying the bounded curvature, given a point set in a plane, and it has attracted lots of attention in the area of robotics. Following the idea of the DTSP, in [5] a smooth path construction (SPC) scheme is proposed to plan paths for a single robot based on a turning circle model. The produced path is smooth. In [39], the limitation of curvature is also considered, and the authors aim to determine a minimum-length circuit whilst maximizing the collection time at each node. In [6], an SVPP strategy is further proposed, which takes into account the collision between sensor nodes and obstacles in the sensing field and the various sensing rates. The authors design trajectories for both the single-robot case and the multiple-robot case, which let the robots avoid hitting sensor nodes and obstacles and be able to collect all the buffered data at sensor nodes.

Most TSP-based approaches fail to consider the features of sensor nodes. In many applications, sensor nodes may have diverse sensing rates, and thus different data loads. If the nodes' storage space is identical, visiting every node with the same frequency leads to buffer overflow for the nodes whose data generation rates are large. Considering this, the paper [40] studies the mobile element scheduling (MES) problem. Different from the TSP framework, one node may be visited multiple times depending on its data generation rate. To address the MES problem, three algorithms, i.e., earliest deadline first (EDF), EDF with k-lookahead, and minimum weight sum first (MWSF) were presented. They extend the work to the case where there are multiple robots to be scheduled in [41]. The authors of [42] also considered the MES problem and proposed a partitioning-based scheduling (PBS) algorithm.

Another interesting work on path design is proposed in [43]. It presents a divide-and-conquer-based path construction method. It discretizes the sensing field into a finite set of candidate turning points. They select some new turning points and add them to the middle of the already constructed path, such that the maximum traffic load from sensor nodes to the robot is minimized. The reference [44] investigates a trajectory optimization problem to maximize the throughput subject to a data collection delay constraint.

The abovementioned approaches focus on the model where all the sensor nodes are visited. This limits the application in large-scale networks. To this end, researchers have proposed cluster-/rendezvous-/anchor-based approaches, where robots only visit the selected sensor nodes, which are promising to significantly shorten the path.

The paper [45] proposes a rendezvous-based approach in which a subset of nodes serves as the rendezvous points (RPs) that buffer data originated from sources and transfer to robots when they arrive. These RPs enable robots to collect a large volume of data at a time without traveling long distances, which can achieve a desirable balance between network energy saving and data collection delay. The authors formulate the minimum-energy rendezvous planning (MERP) problem, which aims to find a set of RPs that can be visited by robots within an allowed delay while the network energy consumed in transmitting data from sources to RPs is minimized. Two rendezvous planning algorithms were developed: RP-CP and RP-UG. RP-CP finds the optimal RPs when MEs move along the data routing tree. RP-UG is a utility-based greedy heuristic that can find RPs with good ratios of network energy saving to ME travel distance. The paper [46] proposes a set packing algorithm and TSP (SPAT) scheme, aiming at guaranteeing data collection from all the sensor nodes in an efficient way. It functions in four phases. Firstly, SPAT generates clusters of sensor nodes. The redundant clusters are removed in the second stage by the set packing algorithm. Thirdly, SPAT adds the least number of clusters required to guarantee that MEs are able to communicate with all the sensor nodes. Finally, SPAT computes the TSP path amongst all of the CHs. The ME moves along this computed path and gathers data from the sensors belonging to various clusters. The aforementioned approaches focus more on the control of robots, while they pay less attention to the routing part. With this consideration, in [47], anchor-based data gathering is considered where an ME periodically starts a data gathering tour, and in each tour it visits some predefined anchor points in the field and stays at each anchor point for a period of sojourn time to collect data from nearby sensors via multihop transmissions, such that the network utility is maximized subject to the constraints on sensor nodes including link capacity, residual energy, and collection delay.

The paper [48] formulates the problem of lifetime maximization as a min-max problem by combining sink mobility together with data routing. They assume that sensors are densely deployed within a circle and there is only one robot responsible for data gathering. They try to find the optimal trajectory for the sink based on analytical models. Since the network density is high, the routing protocol can be simplified and the route for transferring data from any sensor to the robot is a straight line between them. They conclude that the best trajectory of the sink in terms of energy efficiency is to follow the periphery of the network at a constant speed. This work provides theoretical insights into why mobility helps improve the network performance, i.e., the network lifetime. However, to reach the conclusion, several assumptions, such as static traffic, are made. Therefore, the proposed solution, including the trajectory and routing scheme, only has theoretical meaning. In a dense network, communication overheads caused by wireless interference have severe impacts on the network performance. Besides, it is limited in static traffic. In real-world applications, dynamic events lead to dynamic traffic, which could potentially cause unbalanced traffic load in the network when the sink moves at a constant speed on the network periphery. Also, the proposed solution is not scalable, since a larger WSN means a longer periphery, resulting in a significant increment of data latency.

When robots serve as relays, generally, they do not need to move as much as the mobile sinks. The approaches adopting mobile relays generally follow the following system structure: one base station, a set of static sensor nodes, and a set of MEs. Different from the approaches using mobile sinks presented above, or mobile chargers, which will be discussed later, the mobile relays here only move locally and they do not carry data. Besides, the mobile relays are assumed to have richer resource and communication ability than normal static sensor nodes. The basic motivation of employing mobile relays is to relieve sensors that are heavily burdened by high network traffic, thus extending the latter's lifetime, and further the network lifetime.

The paper [49] advocates to use a robot within the vicinity of a base station to take up the role of data routing. The ME can be designed to move around the base station so as to take the role of the static sensors that are running out of power to increase the network lifetime. The merit is that the network lifetime can be enhanced by robots. However, the improvement of network lifetime is very limited since the robot could not eliminate multihop transmission in the network, which is the main factor behind energy consumption of WSNs. The paper [50] considers the cost of robots' movement. The authors show that a robot is allowed to move only when the benefit exceeds the cost and such decision depends on the data packet size. The advantage is that the energy consumed by robots is taken into account. But, it only numerically compares the benefit of robots with its cost because of mobility. The paper [51] extends [50] by designing algorithms and protocols for the collection and distribution of the benefit/cost information, thus enabling local decision making on controlled mobility. The paper [52] considers the problem of where to deploy robots to help certain source nodes transmit the data packets to the static sink taking the packet size into account, such that the total energy consumption of a system consisting of both transmission and robots' movements can be minimized. The authors propose an approach which first computes an optimal routing tree assuming no nodes can move, then improves the topology of the routing tree by greedily adding robots, and finally improves the routing tree by relocating robots (exploiting their mobility) without changing its topology. The proposed model is promising to reduce the energy consumption of data-intensive WSNs. However, the residual power issue has not been considered. The paper [53] claims that the residual energy of sensor nodes is critical in terms of network lifetime. Thus, the authors propose a scheme to calculate the position of a robot, which takes the residual energy of sensor nodes into account.

In [54], the set of sensor nodes form clusters and the CHs are the robots. Each sensor node joins one cluster and reports sensed data to the closest robot. After the clusters are formed, robots start moving within their own clusters and CMs remain the same all the time. Robots collaborate so that each one of them covers approximately the same fraction of the monitored area and cluster areas do not largely overlap. Each robot can either directly communicate with the sink or maintains a connected path to the sink all the time while it is moving. The sensed data are transmitted from the source sensor to a robot and then to the sink. The designed mechanism can be applied to both proactive report and reactive event-driven report. In both proactive report and reactive event-driven report, the data can be transmitted to the sink without a significant delay because of the connectivity, while the requirement that the robots have connectivity all the time results in extra overheads. The paper [55] considers the same problem

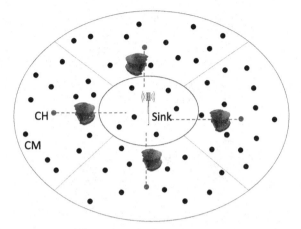

FIGURE 2.4 The system model considered in [55]. Robots can only move between CHs and a position where they can upload data to the sink in one hop.

as [54], but the movement of a robot is restricted to only two positions. The first location is inside a robot's cluster, where the robot can collect data from some nodes within one hop range. The nodes which are within one hop from the first location serve as virtual subsinks for the other nodes in the cluster, i.e., the latter transmit the sensory data to the former, and the former submit the cached data to the robot when it is at the first location. The second location is chosen such that each robot can submit the collected sensory data to the sink in one hop (see, e.g., Fig. 2.4). Based on this structure model, the authors develop an analytical framework to investigate the throughput capacity of the network, defined as the maximal achievable data collection rate from each sensor node. Compared to [54], the connectivity requirement is removed, which makes the approach easy to implement. However, the second assumption, i.e., that each robot can submit the collected data to the sink in one hop, requires robots to have a sufficient communication range; otherwise it prevents the scalability of the proposed scheme.

2.2.1.4 Summary

In this section, we survey the typical approaches of using robots to execute the collection task. According to the mobility model, we categorize these approaches into random, partially controllable, and fully controllable. The random type is easy to be implemented but the system performance is hard to guarantee. Since there is no control over robots, the available approaches mainly focus on the data routing. Therefore, they share most features (multiple robots, no speed control, no trajectory design, energy as the main metric, and distributed solutions), but differ widely in the routing strategy. The fully controllable type can ensure the network performance, while the application area is limited since many environment factors are not considered in the corresponding approaches.

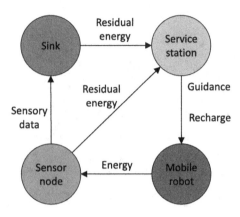

FIGURE 2.5 System structure for energy delivery.

By contrast, the partially controllable type is more suitable in urban environments. This type can be regarded as an extreme case of random type but the predictability is precise. So, the routing strategy receives most research attention. Furthermore, most of these approaches are centralized. Thus, distributed solutions are needed. Beyond the aforementioned publications, some other collection applications have been considered, such as using a mobile sensor network to monitor a set of points of interest (PoIs) [56] or even to discover the PoI [57]. This application requires to relocate the sensors; interested readers can refer to [56,57] and the references therein.

2.2.2 Delivery and combination tasks

In this section, we discuss some relevant publications on the delivery task and the combination task.

The main application in the category of the delivery task includes using robots to charge sensor nodes in a wireless manner. There are also some other applications such as content delivery. For example, robots disseminate information, such as commands or software updates, to sensor nodes. The approaches discussed below are generally based on the following system structure. The system consists of three or four components, namely, one data sink, a set of sensor nodes, one or multiple robots, and/or one service station (SS) (Fig. 2.5). The overall system works as follows. The data sink is the data center and all sensor nodes should transmit their sensory data to it via multihop communication. The robots are responsible to charge the sensor nodes. The SS manages the robots' motions and stores the residual battery levels at sensor nodes. It is also the place where robots can charge their own batteries. In some papers, the SS is omitted, and its functions are added to the sink. In some other papers, robots themselves are embedded with those functions. The information of sensor nodes' residual energy levels can be collected together with the sensory data which are transmitted to the sink first and then submitted to the SS by the sink. In the cases where robots themselves manage such information, routing delivery to robots is adopted.

Basically, the design of using robots to charge sensor nodes involves two main issues:

1. charging schedule, including charging sequence, energy allocation, etc.,

2. routing strategy, i.e., route sensory data to the sink.

2.2.2.1 Nonfully controllable robots

We present some related work falling into the categories of random delivery and partially controllable delivery together in this subsection.

The random robots are rarely employed for the function of delivery. The basic argument is similar to using random mobile sinks for the collection task, i.e., random mobile chargers cannot guarantee a certain system performance. For example, when the network is required to operate for a long time, the sensor nodes are not guaranteed to be charged in time. In [58], a stochastic model is established in which the robot follows a 2D random walk and presents theoretical results on the percentage of sensors that are inactive and suffer from battery failure.

The partially controllable robots also have limitations in this task. The reason lies in the inherent requirement of wireless energy delivery, i.e., the transmitter and receiver should be within proximity. In [59], the authors focus on the scenario where the robot travels along a preplanned trajectory with a time constraint. Due to the fact that the energy received by sensor nodes in a wireless charging manner depends on distance and duration, as well as location diversity, the charged energy at different nodes cannot be maximized simultaneously. Thus, the velocity of the robot plays an important role in energy provision. To this end, the authors try to determine the optimal velocity of the robot subject to a given traveling time constraint, such that the network lifetime is maximized. The paper [60] studies the problem of trajectory selection and node association for robots to minimize unnecessary energy wasted, i.e., movement energy and loss energy, given a set of trajectories and the positions of sensor nodes.

2.2.2.2 Fully controllable robots

Robots can make decisions by themselves or be guided by the SS. Researchers have proposed various schemes to tackle the challenging issues under this scenario, involving the trajectory design for the robot, the sojourn time decision for the robot, and the routing strategy for sensor nodes. Below, we will discuss some representative methods according to their objectives. The involved objective functions include: (1) network lifetime, (2) energy efficiency, (3) charging quantity, and (4) robot number. In the applications where robots are simply to improve the network lifetime, maximizing the network lifetime is usually the priority. When perpetual operation of a WSN is enabled, the research focus turns to the energy efficiency. Here the most widely studied problem is how to efficiently use robots to charge sensor nodes. Furthermore, maximizing the quantity of charging and determining the minimum number of robots to guarantee the continuous operation of WSN also attract much attention.

Network lifetime. As mentioned before, network lifetime is usually the first concern of system designers, especially for long-term WSNs.

The paper [61] considers using a battery capacity-constrained ME to increase the network lifetime. The sensor nodes all have fixed preloaded energy. Due to the limitation of the battery capacity of MEs, the energy added to the network is upper-bounded. Then, how to make use of such precious resource is crucial to improve the network lifetime. The authors claim that the allocation of the finite energy of the ME depends on the energy consumption rates at sensor nodes. To this end, the authors study a joint routing and charging problem to find the optimal routing flow and the amount of energy allocated to each node, subject to the battery capacity constraint of the robot, such that the network lifetime can be maximized. The paper [62] investigates the charging sequence issue. The authors aim at finding an optimal charging sequence together with the corresponding charging time for the sensor nodes to prolong the network lifetime. Compared to [61], this paper considers the issues related to the schedule of the ME, which helps the implementation.

Energy efficiency. The energy consumed by an ME can be classified into three types: movement energy, which is used by robots for moving, loss energy, which varies depending on charging distance and duration, and charged energy, which is eventually received by sensor nodes. Improving the ratio of the received energy to the whole energy consumption increases the energy efficiency of robots. The second class of approaches considers the case that wireless charging is able to make the WSNs operate perpetually. Then, how to improve the energy efficiency of the system becomes an important issue.

The paper [63] points out that the system energy efficiency is fundamentally determined by two factors: the energy consumption rate in the network and the recharging efficiency to the network. Then, the objective of minimizing the charging cost can be achieved by reducing the energy consumption rate and simultaneously improving the charging efficiency. Due to the difficulty of the two-goal coupled problem, the authors propose a sensor deployment scheme to improve the charging efficiency first. With the deployment, they then try to find an optimal routing arrangement and power management at sensor nodes such that the overall data reporting activities can follow the most energy-efficient routes from the source nodes to the sink.

The paper [64] exploits a robot to periodically visit each sensor node and charge it in a wireless manner. Aiming at improving the energy efficiency, routing and charging are jointly considered. Based on the idea in [63], in [65] the scenario is further considered where a post may be deployed with several nodes. Then, the authors propose to divide the field of interest into hexagonal cells and the ME visits the centers of the cells which have sensor nodes inside and charges them. The paper [66] aims at reducing the robot travel distance and the charging delay of sensor nodes simultaneously. Considering the feature that nodes' energy consumption rates are different, only nodes with low remaining energy (on-demand nodes) are involved in the charging round. To this end, a set of nested TSP tours is constructed for the robot.

The paper [67] addresses the shortcoming of [66] by introducing the concept of dynamic charging request. The authors propose a charging discipline called nearest-job-next with preemption, i.e., both charging completion of nodes and the arrival of new charging requests can trigger the reselection of the next to-be-charged node, and the ME selects the spatially closest requesting node at that time as the next node to charge.

The paper [68] proposes a grid-based joint routing and charging algorithm for a single robot containing two phases. In the first phase, a routing protocol is proposed according to the charging characteristics of the ME to balance the energy consumption of sensor nodes locally (within the charging grid). In the second phase, the charging strategy of the robot is designed. Specifically, the strategy includes the path of travel (i.e., traversing each grid point) and charging times at charging points. The goal of designing the charging strategy is to solve the energy unbalance problem caused by the proposed routing protocol, such that the energy balance over the whole network can be achieved globally.

The paper [69] considers the scenario of using robots with limited battery capacity to charge sensor nodes such that the networks can remain operational forever. The onboard battery on a robot is used for both moving and charging. Thus, the number of sensor nodes that one robot can charge is limited, and so is the maximum traveling distance. The authors propose a new paradigm: collaborative charging, i.e., the robot can not only charge sensor nodes but also other robots. Under such paradigm, the problem of scheduling multiple robots is to maximize the ratio of the amount of received energy to movement energy subject to the battery limitation of each robot.

The paper [70] also considers the scenario with dynamic charging requests like [67]. The robots may have various depots and each robot starts to visit sensor nodes from its own depot and finally return to it. Furthermore, the authors formulate an optimization problem whose objective is to minimize the traveling cost of robots, which is closely related to the VRP. The authors design a real-time energy monitoring protocol leveraging the concepts and mechanisms from named data networking (NDN). Compared to the method of transmitting energy together with sensory data, this proposal provides network robustness when intermediate nodes fail.

The paper [71] focuses on a similar objective as [70], i.e., minimizing the total moving distance of robots. But, rather than designing the scheduling for robots only at a specific time instant, in [71] a charging schedule is considered for robots for a given time period. In particular, given the available robots, the charging schedule refers to how many charging rounds should be performed during the time period, which sensor nodes should be charged by which robots in which rounds, and the charging order.

Charging quantity. In many applications with dynamic features, charging all the required sensor nodes is not possible. In this case, many researchers focus on the number of charged sensor nodes, i.e., the charging quantity.

The paper [72] studies the impact of charging based on proximity, like [67]. The authors propose the spatial-dependent task (SDT) scheduling algorithm, which finds a good trade-off between spatial distribution and real-time urgency of charging requests. In [73], the problem of maximization of the charging quantity subject to the traveling time constraint of the ME is studied, in the context of charging on-demand nodes only. The merit is that the practical constraint, i.e., the traveling time of the robot, is taken into account, but it is inferior to [67,72] in terms of charging dynamic on-demand nodes. In [74], the charging time window (which is not considered in [73]) of each to-be-charged node taken into account. They also propose an optimization metric, i.e., charging utility, which takes into account both the fairness of node charging and the charging quantity. Furthermore, the charging time window of each sensor node is considered.

Number of robots. Finding out the minimum number of robots required to achieve an expected system performance is usually the consideration of users, because the system cost depends on the robot number.

The paper [75] considers the scenario of using energy-constrained robots to charge all the sensor nodes in the network. The objective is to find the minimum number of robots. It is assumed that the energy consumption rates at sensor nodes are identical. All the sensor nodes are inclusive in the charging tours. Only the number of robots is the output instead of the charging schedules.

The paper [76] assumes that at a specific time point, the sink receives charging requests and then starts a new routing scheduling by dispatching robots to charge the to-be-charged nodes. Considering the diverse energy consumption rates at sensor nodes [67] and the limited energy capacity at robots [69,75], it aims to find the minimum number of robots. The proposed strategy is only for a specific time point. In [71], on the other hand, the authors focus on a long period of time.

The paper [77] reconsiders the scenario in [69], stating that in [69], the authors only try to minimize the network energy consumption. Since the number of robots impacts the cost of network building, reference [77] studies how to minimize the robot number such that all the sensor nodes can be charged under the battery constraint of robots. With the minimum number of robots, the authors further consider the problem of minimization of the total energy consumption for charging task. They propose the idea of collaborative charging, where robots can charge each other to extend the moving range (see, e.g., Fig. 2.6).

To sum up, we discuss the approaches for the delivery task. Most of them employ fully controllable robots to conduct the task, since the other two mobility types cannot guarantee delivery in time. Generally, the common feature is that the energy consumption rates are different and the energy capacity of robots is

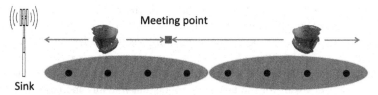

FIGURE 2.6 Illustration of the collaborative charging proposed by [77]. The robot on the left is charged by the sink and it can charge the robot on the right. The meeting point needs to be optimized.

constrained. We categorize them according to the objectives, including network lifetime, energy efficiency, charging quantity, and the robot number. These objectives consider various aspects of WSNs. Compared to the approaches to the collection task, the delivery task has specific requirements. Consider the delivery of energy. Since the efficiency of wireless charging depends highly on the distance between transmitter and receiver, robots need to be in the proximity of sensor nodes to achieve high delivery efficiency. To deliver other content such as information updates, multihop communication can be used like the execution of the collection task. More details will be discussed in Section 2.2.3.1.

2.2.2.3   Combination task

In Sections 2.2.2.1 and 2.2.2.2, when robots are used to charge sensor nodes, they are not responsible for data collection. In this scenario, sensor nodes transmit the sensory data to the sink in similar ways like conventional static WSNs. In contrast, when robots are used to collect data and in the meanwhile charge sensor nodes, the charging strategy may have to be coupled with the routing protocol.

It is possible to simply ignore the routing protocol at sensor nodes, like those in the collection task (e.g., [4,6,39]). A sensor node only transmits its own data to the robot when the robot comes in proximity. Reference [78] is such an example. It forbids multihop routing from sensor nodes to robots and employs robots to collect data from sensor nodes only when they are in proximity and charge the sensor nodes at the same time in the sensing field with various importance levels. In the cases with limited robots, not all sensor nodes are able to operate perpetually. Then, the authors propose to maximize the total weight of the nodes which can work without time limitation. Reducing the problem to the TSP, the authors claim NP-hardness. Furthermore, a greedy algorithm is proposed. Given the locations of the sensor nodes associated with the weights and the charging periods, the greedy algorithm adds the sensor node which has higher weight and lower time requirements to the visiting list of one robot. The algorithm terminates when all the visiting lists of robots are full. In [79], the same scenario is considered as in [78]. Beyond [78], in [79] all sensor nodes are required to be able to operate forever. Adopting the similar greedy algorithm proposed in [78], the minimum number of robots is obtained until all the sensor nodes are added to the robots' visiting lists. The greedy robot scheduling algorithm used in [78,79] is simple, yielding $O(n^2)$, where n is the number of anchor points. However, there is no guarantee for such scheduling. For example, when all the sensor nodes and sinks are on a line, the resulting visiting pattern would be back and forth.

As mentioned before, the scheme of ignoring the routing protocol and letting sensor nodes wait for the arrival of robots unavoidably increases the data collection delay. The papers [80–82] employ a robot to periodically visit some selected sensor positions called anchor points and stay at each anchor point for a sojourn time. The anchor point selection here considers the trade-off between the number of anchor points and the data gathering latency, because from the viewpoint of energy status, more anchor points should be selected to enjoy the benefit of energy replenishment; however, this would adversely prolong the traveling tour length and increase the data gathering latency. The proposed anchor point selection procedure is based on the energy status of sensor nodes, which is collected together with the sensory data. Having that information, a sorted sensor list is constructed, among which the authors try to find a tour whose length is no more than a given threshold. With the sequence to visit anchor points, the authors again formulate the optimization problem with the objective of network utility like in [47].

The paper [83] considers the scenario of using a robot to collect data from the network and charge the sensor nodes in proximity. An optimization problem is formulated, dealing with the moving behavior, i.e., where and how long to make a stop and charge sensor nodes, and how the sensor nodes route the sensory data at each time point. The optimization problem involves variables across multiple dimensions: time, space, and energy. The authors downsize the solution space and transfer the original time-space-dependent problem into a space-dependent subproblem. Moreover, a near-optimal solution to the latter is proposed based on discretization. The idea of downsizing the solution space is a promising way to the time-dependent network problems.

2.2.3 Open issues and future research directions

This section presents an analysis on the relationship between the approaches in different categories and some open issues.

2.2.3.1 Discussion

We have discussed various available schemes for the different roles that robots play in WSNs. The relationship between the approaches within each category has been presented. In this subsection, we go one step further into these approaches and see their similarities across the categories.

The fundamental goal of a WSN is to gather the information of interest, no matter whether robots are used or whether extra energy is added. In conventional static WSNs, the routing strategy plays a critical role. When mobility is taken into account, there are two approaches for data collection: (1) ignore the routing strategy and use only single-hop communication between sensor nodes and robots or (2) develop a mobility-aware routing strategy and use multihop communication. For simplicity, consider a single-robot case. The data collection delay in the first approach would be the tour time of the robot, while in the second approach it would be the time the robot spends on traveling between two locations. The saved delay comes with the cost of a large amount of communication overhead to announce the position of the robot. When wireless charging is taken into consideration, there are also two approaches for data collection: (3) the data center remains at the static base station and the robot is only used to charge sensor nodes or (4) the charging task and the collection task are combined at the robot. The differences between these approaches are quite clear; now we discuss some similarities between them:

• When wireless charging is considered, all the sensor nodes should be visited sooner or later to make the network survive forever. From this viewpoint, both (3) and (4) need to charge sensor nodes in proximity, which is another single-hop communication between sensor nodes and robots. The difference lies in the content direction: in (1) data packets are sent to robots by sensor nodes, while in (3) and (4) energy is sent to sensor nodes by robots.

• When sensor nodes have various energy consumption rates, not all sensor nodes are required to be charged in the same robot tours, i.e., robots only need to charge on-demand nodes. In this scenario, one similarity of (2) and (4) is that they both need to consider the trajectory design and the mobility-aware routing strategy.

• More specifically, in many realistic applications, the wireless data transmission time should be considered, especially in cases of large data packets, such as video. Besides, the experimental results show that the wireless charging of a sensor node to full battery capacity usually needs hours [62]. Both the wireless data transmission time and the wireless charging time may influence the decision on the sojourn time a robot spends at a sensor node in (1) and (3). In particular, the larger the packet, the more sojourn time is required in (1), while the less the residual energy at a node, the more sojourn time is required in (3).

From the above analysis, it becomes clear that collection and delivery are two opposite functions in some circumstances.

2.2.3.2   Open issues

This subsection highlights some open topics in the area of using robots in WSNs.

3D WSNs. Most of the available approaches consider the case where sensor nodes are deployed on a 2D plane and usually ground robots or vehicles are used as mobile platforms to carry robots. There are also some publications about the use of UAVs to execute the data collection task [5,44]. The UAVs do not change altitude, which suggests it is in essence still a 2D problem. References [84,85] consider the problem of using UAVs to monitor ground targets, and the UAVs can be regarded as mobile sensor nodes in 3D space. In some applications, such as crucial infrastructure monitoring [86], the static sensor nodes are deployed on the surface or inside the infrastructures. Another scenario is the underground or underwater data collection and/or charging system [87], where the sensor nodes are also deployed in 3D environments. In these cases, the 2D-based approaches are not sufficient.

Thus, it would be necessary to design schemes suitable for 3D WSNs. The approaches for 3D WSNs would be more complex than the existing ones for 2D WSNs, and those for the applications inside buildings would be more challenging than the underwater scenarios, since the environment inside buildings may be more cluttered, resulting in more constraints on robot movement.

Charging mobile sensors. In this survey, we mostly focus on the scenario with static sensor nodes. In some applications, sensor nodes may need to track targets and gather data from the targets [88]. In these cases, normal sensor nodes can be mobile. This feature also leads to many other problems such as the coverage of the area of interest, including barrier coverage, blanket coverage, and sweep coverage [89–92]. When the sensor nodes execute the tracking task, the sensors' movements must be carefully considered, since it relates to the energy efficiency of the system and network lifetime [93]. Some aspects such as the connectivity of the sensor nodes need to be taken into account as well when they are mobile. Furthermore, to the best of our knowledge, the currently available energy delivery approaches all focus on the scenario with static sensor nodes, and cannot be applied directly to the cases with mobile sensor nodes. Though harvesting energy by mobile sensor nodes is a promising direction [94], another research direction is to make use of robots to charge mobile sensor nodes in a wireless manner such that the network can operate for a long time, which will be less impacted by the environment than the case where energy harvesting is adopted.

Sensor-actuator networks with robots. Another interesting direction of future research is to study the use of robots in sensor-actuator networks. In such networks, some nodes are sensors and some are actuators (also called actors), whereas other nodes are endowed with both sensing and actuating capacities. Actuating capabilities are utilized to dispense control signals with the goal of achieving certain control objectives. Moreover, some nodes of such wireless networks may be mobile robots. Many modern engineering applications include the use of such networks to provide efficient and effective monitoring and control of industrial and environmental processes. These networks are able to achieve improved performance, along with a reduction in power consumption and production cost. An important open area of research is the study of coverage control problems for wireless sensor/actuator networks. In particular, one open problem is the novel problem of termination of a moving unknown environmental region by a sensor-actuator network with some mobile robotic nodes introduced in [95]. In real-world applications, this moving region may represent an oil spill or an area contaminated with a hazardous chemical or biological agent. In this problem, we assume that part of the nodes in the network are mobile autonomous robots. Furthermore, they are equipped with not only sensors but also actuators that release a neutralizing chemical to control the shape of the polluted region. In other words, some nodes of the network are capable to not just measure the moving region in their neighborhoods, but also to terminate parts of this region. Moreover, in some problems, the field can terminate the sensors/actuators as well. In such applications, actuation plays a major role. The goal is to achieve the complete termination of moving hazardous fields in realistic situations. In particular, a challenging open problem is to develop termination algorithms with guaranteed termination of the moving region under certain assumptions on the unknown environmental field in a finite time. Moreover, the time-optimal termination problem for various classes of time-varying environmental fields should be studied. In other words, the problem here is to find a decentralized strategy for the wireless sensor/actuator network that achieves termination in the minimum possible time. From a control-theoretic viewpoint some theoretical results on the control of sensor/actuator networks were obtained in [96], but these results are still far from application in real engineering problems.

2.3 Concluding remarks

Enabling an extra dimension by mobility improves the performance of WSNs in various aspects. This chapter gives a comprehensive survey of existing typical approaches following the taxonomy based on the tasks executed by robots, i.e., collection, delivery, and combination. Besides summaries on the existing studies, potential extensions have also been discussed. From the survey, we can conclude that the tasks of collection and delivery are opposite in certain circumstances. Thus, some research activities on these two tasks can be merged in the future. More importantly, there are still a few open issues that have attracted little attention so far. Some basic ideas to address them by combining the strategies used in related areas have been discussed.

References

[1] L. Atzori, A. Iera, G. Morabito, The internet of things: a survey, Computer Networks 54 (15) (2010) 2787–2805.

[2] I.F. Akyildiz, W. Su, Y. Sankarasubramaniam, E. Cayirci, Wireless sensor networks: a survey, Computer Networks 38 (4) (2002) 393–422.

[3] H. Huang, A.V. Savkin, M. Ding, C. Huang, Mobile robots in wireless sensor networks: a survey on tasks, Computer Networks 148 (2019) 1–19.

[4] R.C. Shah, S. Roy, S. Jain, W. Brunette, Data mules: modeling and analysis of a three-tier architecture for sparse sensor networks, Ad Hoc Networks 1 (2) (2003) 215–233.

[5] A. Wichmann, T. Korkmaz, Smooth path construction and adjustment for multiple mobile sinks in wireless sensor networks, Computer Communications 72 (2015) 93–106.

[6] H. Huang, A.V. Savkin, Viable path planning for data collection robots in a sensing field with obstacles, Computer Communications 111 (2017) 84–96.

[7] F. Ye, H. Luo, J. Cheng, S. Lu, L. Zhang, A two-tier data dissemination model for large-scale wireless sensor networks, in: MobiCom, ACM, 2002, pp. 148–159.

[8] C.-J. Lin, P.-L. Chou, C.-F. Chou, HCDD: hierarchical cluster-based data dissemination in wireless sensor networks with mobile sink, in: International Conference on Wireless Communications and Mobile Computing, ACM, 2006, pp. 1189–1194.

[9] X. Liu, H. Zhao, X. Yang, X. Li, Sinktrail: a proactive data reporting protocol for wireless sensor networks, IEEE Transactions on Computers 62 (1) (2013) 151–162.

[10] D. Kim, R. Uma, B.H. Abay, W. Wu, W. Wang, A.O. Tokuta, Minimum latency multiple data mule trajectory planning in wireless sensor networks, IEEE Transactions on Mobile Computing 13 (4) (2014) 838–851.

[11] M. Vecchio, A.C. Viana, A. Ziviani, R. Friedman, DEEP: density-based proactive data dissemination protocol for wireless sensor networks with uncontrolled sink mobility, Computer Communications 33 (8) (2010) 929–939.

[12] E.B. Hamida, G. Chelius, A line-based data dissemination protocol for wireless sensor networks with mobile sink, in: ICC, IEEE, 2008, pp. 2201–2205.

[13] P. Baruah, R. Urgaonkar, B. Krishnamachari, Learning-enforced time domain routing to mobile sinks in wireless sensor fields, in: LCN, IEEE, 2004, pp. 525–532.

[14] H. Lee, M. Wicke, B. Kusy, O. Gnawali, L. Guibas, Predictive data delivery to mobile users through mobility learning in wireless sensor networks, IEEE Transactions on Vehicular Technology 64 (12) (2015) 5831–5849.

[15] I. Papadimitriou, L. Georgiadis, Maximum lifetime routing to mobile sink in wireless sensor networks, in: SoftCOM, 2005.

[16] S.R. Gandham, M. Dawande, R. Prakash, S. Venkatesan, Energy efficient schemes for wireless sensor networks with multiple mobile base stations, in: GLOBECOM, vol. 1, IEEE, 2003, pp. 377–381.

[17] S. Basagni, A. Carosi, E. Melachrinoudis, C. Petrioli, Z.M. Wang, Controlled sink mobility for prolonging wireless sensor networks lifetime, Wireless Networks 14 (6) (2008) 831–858.

[18] S. Basagni, A. Carosi, C. Petrioli, C.A. Phillips, Moving multiple sinks through wireless sensor networks for lifetime maximization, in: MASS, IEEE, 2008, pp. 523–526.

[19] Y. Yun, Y. Xia, Maximizing the lifetime of wireless sensor networks with mobile sink in delay-tolerant applications, IEEE Transactions on Mobile Computing 9 (9) (2010) 1308–1318.

[20] Y. Gu, Y. Ji, J. Li, B. Zhao, ESWC: efficient scheduling for the mobile sink in wireless sensor networks with delay constraint, IEEE Transactions on Parallel and Distributed Systems 24 (7) (2013) 1310–1320.

[21] Y. Shi, Y.T. Hou, Theoretical results on base station movement problem for sensor network, in: INFOCOM, IEEE, 2008, pp. 1–5.

[22] A. Chakrabarti, A. Sabharwal, B. Aazhang, Using predictable observer mobility for power efficient design of sensor networks, in: Information Processing in Sensor Networks, Springer, 2003, pp. 129–145.

[23] L. Song, D. Hatzinakos, Architecture of wireless sensor networks with mobile sinks: sparsely deployed sensors, IEEE Transactions on Vehicular Technology 56 (4) (2007) 1826–1836.

[24] A. Mehrabi, K. Kim, Maximizing data collection throughput on a path in energy harvesting sensor networks using a mobile sink, IEEE Transactions on Mobile Computing 15 (3) (2016) 690–704.

[25] A.A. Somasundara, A. Kansal, D.D. Jea, D. Estrin, M.B. Srivastava, Controllably mobile infrastructure for low energy embedded networks, IEEE Transactions on Mobile Computing 5 (8) (2006) 958–973.

[26] S. Gao, H. Zhang, S.K. Das, Efficient data collection in wireless sensor networks with path-constrained mobile sinks, IEEE Transactions on Mobile Computing 10 (4) (2011) 592–608.

[27] C. Konstantopoulos, G. Pantziou, D. Gavalas, A. Mpitziopoulos, B. Mamalis, A rendezvous-based approach enabling energy-efficient sensory data collection with mobile sinks, IEEE Transactions on Parallel and Distributed Systems 23 (5) (2012) 809–817.

[28] G. Chen, C. Li, M. Ye, J. Wu, An unequal cluster-based routing protocol in wireless sensor networks, Wireless Networks 15 (2) (2009) 193–207.

[29] H. Huang, A.V. Savkin, An energy efficient approach for data collection in wireless sensor networks using public transportation vehicles, AEÜ. International Journal of Electronics and Communications 75 (2017) 108–118.

[30] H. Huang, A.V. Savkin, C. Huang, I-UMDPC: the improved-unusual message delivery path construction for wireless sensor networks with mobile sinks, IEEE Internet of Things Journal 4 (5) (2017) 1528–1536.

[31] E.L. Lawler, J.K. Lenstra, A.R. Kan, D.B. Shmoys, et al., The Traveling Salesman Problem: a Guided Tour of Combinatorial Optimization, vol. 3, Wiley, New York, 1985.

[32] A. Dumitrescu, J.S. Mitchell, Approximation algorithms for TSP with neighborhoods in the plane, in: The 12th Annual ACM-SIAM Symposium on Discrete Algorithms, Society for Industrial and Applied Mathematics, 2001, pp. 38–46.

[33] B. Yuan, M. Orlowska, S. Sadiq, On the optimal robot routing problem in wireless sensor networks, IEEE Transactions on Knowledge and Data Engineering 19 (9) (2007) 1252–1261.

[34] R. Sugihara, R.K. Gupta, Optimizing Energy-Latency Trade-Off in Sensor Networks with Controlled Mobility, IEEE, 2009.

[35] L. He, J. Pan, J. Xu, A progressive approach to reducing data collection latency in wireless sensor networks with mobile elements, IEEE Transactions on Mobile Computing 12 (7) (2013) 1308–1320.

[36] G. Laporte, The vehicle routing problem: an overview of exact and approximate algorithms, European Journal of Operational Research 59 (3) (1992) 345–358.

[37] L.E. Dubins, On curves of minimal length with a constraint on average curvature, and with prescribed initial and terminal positions and tangents, American Journal of Mathematics 79 (3) (1957) 497–516.

[38] K. Savla, E. Frazzoli, F. Bullo, On the point-to-point and traveling salesperson problems for Dubins' vehicle, in: American Control Conference, IEEE, 2005, pp. 786–791.

[39] D.G. Macharet, J.W. Monteiro, G.R. Mateus, M.F. Campos, Bi-objective data gathering path planning for vehicles with bounded curvature, Computers & Operations Research 84 (2017) 195–204.

[40] A. Somasundara, A. Ramamoorthy, M.B. Srivastava, et al., Mobile element scheduling for efficient data collection in wireless sensor networks with dynamic deadlines, in: RTSS, IEEE, 2004, pp. 296–305.

[41] A. Somasundara, A. Ramamoorthy, M.B. Srivastava, et al., Mobile element scheduling with dynamic deadlines, IEEE Transactions on Mobile Computing 6 (4) (2007) 395–410.

[42] Y. Gu, D. Bozdag, E. Ekici, F. Özgüner, C.-G. Lee, Partitioning based mobile element scheduling in wireless sensor networks, in: SECON, Citeseer, 2005, pp. 386–395.

[43] M. Ma, Y. Yang, SenCar: an energy-efficient data gathering mechanism for large-scale multi-hop sensor networks, IEEE Transactions on Parallel and Distributed Systems 18 (10) (2007) 1476–1488.

[44] L. Chen, W. Wang, H. Huang, S. Lin, On time-constrained data harvesting in wireless sensor networks: approximation algorithm design, IEEE/ACM Transactions on Networking 24 (5) (2016) 3123–3135.

[45] G. Xing, T. Wang, Z. Xie, W. Jia, Rendezvous planning in wireless sensor networks with mobile elements, IEEE Transactions on Mobile Computing 7 (12) (2008) 1430–1443.

[46] H. Nakayama, Z.M. Fadlullah, N. Ansari, N. Kato, A novel scheme for WSAN sink mobility based on clustering and set packing techniques, IEEE Transactions on Automatic Control 56 (10) (2011) 2381–2389.

[47] M. Zhao, Y. Yang, Optimization-based distributed algorithms for mobile data gathering in wireless sensor networks, IEEE Transactions on Mobile Computing 11 (10) (2012) 1464–1477.

[48] J. Luo, J.-P. Hubaux, Joint mobility and routing for lifetime elongation in wireless sensor networks, in: INFOCOM, vol. 3, IEEE, 2005, pp. 1735–1746.

[49] W. Wang, V. Srinivasan, K.-C. Chua, Extending the lifetime of wireless sensor networks through mobile relays, IEEE/ACM Transactions on Networking 16 (5) (2008) 1108–1120.

[50] D.K. Goldenberg, J. Lin, A.S. Morse, B.E. Rosen, Y.R. Yang, Towards mobility as a network control primitive, in: MobiHoc, ACM, 2004, pp. 163–174.

[51] C. Tang, P.K. McKinley, Energy optimization under informed mobility, IEEE Transactions on Parallel and Distributed Systems 17 (9) (2006) 947–962.

[52] F. El-Moukaddem, E. Torng, G. Xing, Mobile relay configuration in data-intensive wireless sensor networks, IEEE Transactions on Mobile Computing 12 (2) (2013) 261–273.

[53] F.-N. Huang, T.-Y. Chen, S.-H. Chen, H.-W. Wei, T.-s. Hsu, W.-K. Shih, Shareenergy: configuring mobile relays to extend sensor networks lifetime based on residual power, in: International Conference on Collaboration Technologies and Systems (CTS), IEEE, 2016, pp. 478–484.

[54] T. Banerjee, B. Xie, J.H. Jun, D.P. Agrawal, Increasing lifetime of wireless sensor networks using controllable mobile cluster heads, Wireless Communications and Mobile Computing 10 (3) (2010) 313–336.

[55] W. Liu, K. Lu, J. Wang, L. Huang, D.O. Wu, On the throughput capacity of wireless sensor networks with mobile relays, IEEE Transactions on Vehicular Technology 61 (4) (2012) 1801–1809.

[56] A. Trotta, F.D. Andreagiovanni, M. Di Felice, E. Natalizio, K.R. Chowdhury, When UAVs ride a bus: towards energy-efficient city-scale video surveillance, in: INFOCOM, IEEE, 2018, pp. 1043–1051.

[57] T. Razafindralambo, M. Erdelj, D. Zorbas, E. Natalizio, Spread and shrink: point of interest discovery and coverage with mobile wireless sensors, Journal of Parallel and Distributed Computing 102 (2017) 16–27.

[58] C. Wang, Y. Yang, J. Li, Stochastic mobile energy replenishment and adaptive sensor activation for perpetual wireless rechargeable sensor networks, in: Wireless Communications and Networking Conference (WCNC), IEEE, 2013, pp. 974–979.

[59] Y. Shu, H. Yousefi, P. Cheng, J. Chen, Y.J. Gu, T. He, K.G. Shin, Near-optimal velocity control for mobile charging in wireless rechargeable sensor networks, IEEE Transactions on Mobile Computing 15 (7) (2016) 1699–1713.

[60] S. Zhang, Z. Qian, J. Wu, F. Kong, S. Lu, Optimizing itinerary selection and charging association for mobile chargers, IEEE Transactions on Mobile Computing 16 (10) (2017) 2833–2846.

[61] Z. Li, Y. Peng, W. Zhang, D. Qiao, Study of joint routing and wireless charging strategies in sensor networks, in: WASA, Springer, 2010, pp. 125–135.

[62] Y. Peng, Z. Li, W. Zhang, D. Qiao, Prolonging sensor network lifetime through wireless charging, in: RTSS, IEEE, 2010, pp. 129–139.

[63] B. Tong, Z. Li, G. Wang, W. Zhang, How wireless power charging technology affects sensor network deployment and routing, in: The 30th International Conference on Distributed Computing Systems (ICDCS), IEEE, 2010, pp. 438–447.

[64] Y. Shi, L. Xie, Y.T. Hou, H.D. Sherali, On renewable sensor networks with wireless energy transfer, in: INFOCOM, IEEE, 2011, pp. 1350–1358.

[65] L. Xie, Y. Shi, Y.T. Hou, W. Lou, H.D. Sherali, S.F. Midkiff, On renewable sensor networks with wireless energy transfer: the multi-node case, in: SECON, IEEE, 2012, pp. 10–18.

[66] L. He, L. Fu, L. Zheng, Y. Gu, P. Cheng, J. Chen, J. Pan, Esync: an energy synchronized charging protocol for rechargeable wireless sensor networks, in: MobiHoc, ACM, 2014, pp. 247–256.

[67] L. He, L. Kong, Y. Gu, J. Pan, T. Zhu, Evaluating the on-demand mobile charging in wireless sensor networks, IEEE Transactions on Mobile Computing 14 (9) (2015) 1861–1875.

[68] G. Han, A. Qian, J. Jiang, N. Sun, L. Liu, A grid-based joint routing and charging algorithm for industrial wireless rechargeable sensor networks, Computer Networks 101 (2016) 19–28.

[69] S. Zhang, J. Wu, S. Lu, Collaborative mobile charging for sensor networks, in: MASS, IEEE, 2012, pp. 84–92.

[70] C. Wang, J. Li, F. Ye, Y. Yang, NETWRAP: an NBN based real-time wireless recharging framework for wireless sensor networks, IEEE Transactions on Mobile Computing 13 (6) (2014) 1283–1297.

[71] W. Xu, W. Liang, X. Lin, G. Mao, Efficient scheduling of multiple mobile chargers for wireless sensor networks, IEEE Transactions on Vehicular Technology 65 (9) (2016) 7670–7683.

[72] H. Huang, S. Lin, L. Chen, J. Gao, A. Mamat, J. Wu, Dynamic mobile charger scheduling in heterogeneous wireless sensor networks, in: MASS, IEEE, 2015, pp. 379–387.

[73] X. Ren, W. Liang, W. Xu, Maximizing charging throughput in rechargeable sensor networks, in: The 23rd International Conference on Computer Communication and Networks (ICCCN), IEEE, 2014, pp. 1–8.

[74] X. Ye, W. Liang, Charging utility maximization in wireless rechargeable sensor networks, Wireless Networks 23 (7) (2017) 2069–2081.

[75] H. Dai, X. Wu, G. Chen, L. Xu, S. Lin, Minimizing the number of mobile chargers for large-scale wireless rechargeable sensor networks, Computer Communications 46 (2014) 54–65.

[76] W. Liang, W. Xu, X. Ren, X. Jia, X. Lin, Maintaining sensor networks perpetually via wireless recharging mobile vehicles, in: LCN, IEEE, 2014, pp. 270–278.

[77] T. Liu, B. Wu, H. Wu, J. Peng, Low-cost collaborative mobile charging for large-scale wireless sensor networks, IEEE Transactions on Mobile Computing 16 (8) (2017) 2213–2227.

[78] K.L.-M. Ang, J.K.P. Seng, A.M. Zungeru, Optimizing energy consumption for big data collection in large-scale wireless sensor networks with mobile collectors, IEEE Systems Journal 12 (1) (2018) 616–626.

[79] B.-H. Liu, N.-T. Nguyen, V.-T. Pham, Y.-X. Lin, Novel methods for energy charging and data collection in wireless rechargeable sensor networks, International Journal of Communication Systems 30 (5) (2017).

[80] J. Li, M. Zhao, Y. Yang, OWER-MDG: a novel energy replenishment and data gathering mechanism in wireless rechargeable sensor networks, in: GLOBECOM, IEEE, 2012, pp. 5350–5355.

[81] S. Guo, C. Wang, Y. Yang, Joint mobile data gathering and energy provisioning in wireless rechargeable sensor networks, IEEE Transactions on Mobile Computing 13 (12) (2014) 2836–2852.

[82] M. Zhao, J. Li, Y. Yang, A framework of joint mobile energy replenishment and data gathering in wireless rechargeable sensor networks, IEEE Transactions on Mobile Computing 13 (12) (2014) 2689–2705.

[83] L. Xie, Y. Shi, Y.T. Hou, W. Lou, H.D. Sherali, On traveling path and related problems for a mobile station in a rechargeable sensor network, in: MobiHoc, ACM, 2013, pp. 109–118.

[84] L. Di Puglia Pugliese, F. Guerriero, D. Zorbas, T. Razafindralambo, Modelling the mobile target covering problem using flying drones, Optimization Letters 10 (Jun 2016) 1021–1052.

[85] D. Zorbas, L.D.P. Pugliese, T. Razafindralambo, F. Guerriero, Optimal drone placement and cost-efficient target coverage, Journal of Network and Computer Applications 75 (2016) 16–31.

[86] S. Kim, S. Pakzad, D. Culler, J. Demmel, G. Fenves, S. Glaser, M. Turon, Health monitoring of civil infrastructures using wireless sensor networks, in: The 6th International Conference on Information Processing in Sensor Networks, ACM, 2007, pp. 254–263.

[87] L. Liu, M. Ma, C. Liu, Y. Shu, Optimal relay node placement and flow allocation in underwater acoustic sensor networks, IEEE Transactions on Communications 65 (5) (2017) 2141–2152.

[88] C. Zhu, L. Shu, T. Hara, L. Wang, S. Nishio, L.T. Yang, A survey on communication and data management issues in mobile sensor networks, Wireless Communications and Mobile Computing 14 (1) (2014) 19–36.

[89] T.M. Cheng, A.V. Savkin, A distributed self-deployment algorithm for the coverage of mobile wireless sensor networks, IEEE Communications Letters 13 (11) (2009).

[90] A.V. Savkin, F. Javed, A.S. Matveev, Optimal distributed blanket coverage self-deployment of mobile wireless sensor networks, IEEE Communications Letters 16 (6) (2012) 949–951.

[91] T.M. Cheng, A.V. Savkin, Decentralized control for mobile robotic sensor network self-deployment: barrier and sweep coverage problems, Robotica 29 (2) (2011) 283–294.

[92] S. He, J. Chen, X. Li, X.S. Shen, Y. Sun, Mobility and intruder prior information improving the barrier coverage of sparse sensor networks, IEEE Transactions on Mobile Computing 13 (6) (2014) 1268–1282.

[93] Z. Liao, J. Wang, S. Zhang, J. Cao, G. Min, Minimizing movement for target coverage and network connectivity in mobile sensor networks, IEEE Transactions on Parallel and Distributed Systems 26 (July 2015) 1971–1983.

[94] D. Zorbas, P. Raveneau, Y. Ghamri-Doudane, Assessing the cost of deploying and maintaining indoor wireless sensor networks with RF-power harvesting properties, Pervasive and Mobile Computing 43 (2018) 64–77.

[95] A.V. Savkin, T.M. Cheng, Z. Xi, F. Javed, A.S. Matveev, H. Nguyen, Decentralized Coverage Control Problems for Mobile Robotic Sensor and Actuator Networks, IEEE Press—Wiley, 2015.

[96] M.A. Demetriou, Guidance of mobile actuator-plus-sensor networks for improved control and estimation of distributed parameter systems, IEEE Transactions on Automatic Control 55 (7) (2010) 1570–1584.

Chapter 3

Wireless communication networks supported by autonomous UAVs: a survey[☆]

3.1 Introduction

UAVs have been widely applied in civilian and commercial applications, which include but are not limited to wireless communication, agriculture, logistics, security, search and rescue, remote sensing, and entertainment [1–7]. Among these applications, the usage of UAVs in wireless communication networks has attracted great attention from researchers and practitioners. Thanks to the improved mobility and reduced cost, UAVs are regarded as an important tool to reshape the future wireless communication networks.

Firstly, UAVs can serve cellular users. They can play the role of aerial base stations to provide communication services to cellular users, especially in some congested urban areas [8]. This is a promising solution to 5G or beyond-5G networks. Also, UAVs can service WSNs [9]. Working as flying data sinks, these UAVs can collect sensory data from distributed and/or sparsely deployed sensor nodes; and working as chargers, they can recharge sensor nodes' batteries in a wireless manner, in order to prolong the network lifetime. Moreover, they can navigate ground robots since they may have a better view of the environment, and they can also collaborate with ground robots to execute complex tasks.

Despite having many promising opportunities, several key issues must be addressed to facilitate the usage of UAVs. The first issue is coverage. Generally, the system providers wish to provide a system that has sufficient coverage of targets, either humans or field sites. The second issue is connectivity. Specifically, forming a connected network enables fast information transmission between UAVs and the ground control stations. The third issue is about energy. Since most commercial UAVs currently available on the market are powered by batteries, their working time is limited. Thus, how to ensure a persistent service is another aspect of quality of service.

In this chapter, we review the available approaches relevant to the applications of UAVs in wireless communication networks. Based on our review, some

---

☆ The main material of this chapter was originally published in Hailong Huang, Andrey V. Savkin, Towards the internet of flying robots: a survey, Sensors 18 (11) (2018) 4038.

https://doi.org/10.1016/B978-0-32-390182-6.00008-2
Copyright © 2022 Elsevier Inc. All rights reserved.

comments on the existing approaches are further provided, their limitations are pointed out, and some promising future research directions are suggested. The main material of this chapter was originally published in [10].

The rest of the chapter is organized as follows. Section 3.2 reviews approaches related to applications involving human users. In particular, this section focuses on common issues including coverage, connectivity, and energy in these applications. Section 3.3 reviews approaches for the UAVs collaborating with WSNs. Section 3.4 presents an overview of the limitations of existing approaches, followed by some open issues which have not received much attention from researchers. Finally, Section 3.5 briefly summarizes the chapter.

3.2 UAVs serve humans and WSNs

This section discusses some common issues when UAVs serve humans. In particular, some basic models used in different applications are presented first, and then typical issues from the optimization point of view are discussed.

3.2.1 Coverage model

Coverage is usually the first issue the system designers need to consider. Two popular coverage models are observed in the literature: the disk model and the Voronoi cell model. This classification is due to different system assumptions. Specifically, when interference is not considered in the system, a disk coverage is adopted; while when interference is taken into account, a more precise mechanism is necessary to characterize the coverage.

3.2.1.1 Disk coverage

The disk model is the simplest coverage model. Similar to the down facing camera coverage [9,11,12], the disk model can be characterized by a coverage radius. In the applications of providing wireless communication service to cellular users, the coverage of a user by a UAV can be determined by the signal path loss (PL), under the assumption that the user and the UAV have LoS. References [13,14] consider that the links between UAVs and ground users can have two cases: LoS and non-LoS (NLoS). The authors of [14] propose a model for the probability of having an LoS link between the two parts (P_{LoS}), which depends on the elevation angle and some environmental parameters:

$$P_{LoS} = \frac{1}{1 + a\exp(-b(\varphi - a))}, \tag{3.1}$$

where φ is the elevation angle (Fig. 3.1) and a and b are environment-dependent parameters. As pointed out by [14], a and b depend on environmental parameters including the ratio of built-up land area to the total land area, the mean number of buildings per unit area, and a scale parameter that describes the buildings' height distribution according to a Rayleigh probability density function. The

Flying robot

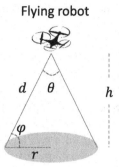

FIGURE 3.1 Disk coverage model.

probability of NLoS link is $P_{NLoS} = 1 - P_{LoS}$. Furthermore, the PL is modeled with two parts: free space PL and excessive PL η_ξ, where $\xi \in \{LoS, NLoS\}$. Free space PL depends on the distance between the UAV and the ground user, while excessive PL depends on the type of link between the two parts. Thus, the average PL from the UAV to the ground user is the sum of the LoS PL and NLoS PL [14]:

$$PL = P_{LoS} PL_{LoS} + P_{NLoS} PL_{NLoS}, \qquad (3.2)$$

where $PL_\xi = 20\log(\frac{4\pi f d}{c}) + \eta_\xi$, d is the Euclidean distance between the UAV and the ground user, f is the carrier frequency, and c is the speed of light. Furthermore, by setting a maximum allowed PL, one can compute the largest coverage radius for a given altitude. Furthermore, the optimal altitude h, which corresponds to the global largest coverage radius, can also be obtained [14].

The disk coverage model is also suitable for applications of wirelessly charging sensor nodes by UAVs [15] under the consumption of LoS. Unlike many other power harvesting methods, such as solar and vibration, radio frequency (RF)-power harvesting can recharge multiple devices simultaneously, and it is not significantly dependent on the environment. However, the received power and the efficiency of the harvesting module of RF-power harvesting are both highly dependent on the distance between the charger and node. In [15], the energy harvesting efficiency by a node depends on two terms: the received power and the efficiency of the harvesting antenna. Both of them depend on the distance between the charger and the node. For the former, a prorogation model proposed in [16] is adopted, where the received power decreases with increasing distance. For the latter, the efficiency values provided by the manufacturer Powercast [17] are used. Consider a scenario where a UAV can charge a ground sensor node for a limited period. Taking into account the sensor nodes' energy consumption model and setting the objective as fully replenishing the nodes' battery, one can obtain the maximum distance d_{max} between a flying charger and a sensor node [15]. For a given UAV at some position and under the above setting, a node can work without a time limit if it is within d_{max} of a flying charger.

3.2.1.2 Voronoi cell model

One implied assumption of the disk model is that only one UAV is used to serve ground users, or multiple UAVs use different wireless channels. Regarding this, a more practical communication coverage model should take interference into account. In the above model, the coverage of a user or the association of it with a UAV is determined by the maximum allowed PL or signal-to-noise ratio (SNR). When interference is taken into account, the coverage or association should be determined by the signal-to-interference-and-noise ratio (SINR). Following the average PL model [14], it can be seen that the received signal from the nearest UAV is always the largest. Thus, by assuming that the association of a user and a UAV is determined by SINR, a user is always associated with the nearest UAV [18]. Obviously, the area of interest can be separated into several Voronoi cells given the positions of UAVs (Fig. 3.2).

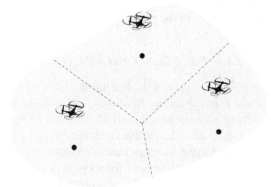

FIGURE 3.2 Coverage areas based on Voronoi cells.

3.2.2 Connectivity

Many applications of wireless communication networks involving UAVs require that they form a connected network. For example, in the application of data collection from wireless nodes using UAVs, the connectivity of UAVs with the central data sink guarantees that sensory data can be delivered to the data sink within a reasonable delay, which makes it possible that end-users take necessary actions timely. The case is the same in data dissemination. Another application is using UAVs to provide communication services to ground cellular users. The connectivity of UAVs with stationary base stations (SBSs) ensures that every UAV has a wireless backhaul link so that any request from users can be transmitted to the core networks instantly and the response can also be returned to the user shortly. The connectivity requirement significantly influences the deployment of UAVs, especially in disaster areas. In such areas, all or most of the existing SBSs may be destroyed by the disaster; thus, the UAVs need to construct a new communication system and connect themselves to the remote working SBSs.

To this end, one simple model to characterize the connectivity requirement has been proposed in [19]. Consider a communication system consisting of n UAVs and m SBSs, and the UAVs are working at the same altitude. Let $P_1, P_2, \ldots, P_n$ be the coordinates of UAVs on the horizontal plane and let $Q_1, Q_2, \ldots, Q_m$ be the fixed locations of SBSs. The connectivity of such a communication system can be described by a communication graph $\mathcal{G}$ [19]. In the given communication graph $\mathcal{G}$, there are $n + m$ vertices and any robot vertex should be connected to an SBS vertex. The connectivity of two robot vertices and one robot vertex and one SBS vertex can be described by

$$D(P_i, P_j) \leq R_1 \tag{3.3}$$

if robot vertices i and j are connected by an edge in $\mathcal{G}$, and by

$$D(P_i, Q_j) \leq R_2 \tag{3.4}$$

if robot vertex i and SBS vertex j are connected by an edge in $\mathcal{G}$, where $D(\cdot, \cdot)$ denotes the 2D distance between two vertices and R_1 and R_2 are given constants. Based on the idea of [19], 3D connectivity can be obtained easily by introducing the altitude dimension [9,20].

Besides the connectivity discussed here, which focuses on the UAVs and SBSs, another concept relating to connectivity is that UAVs can work as relays to link disconnected networks. Interested readers are referred to [21] and the references therein.

3.2.3   Energy consumption

It is clear that many current commercial UAVs are powered by onboard batteries, which means that their working time is limited. Some publications focus on flight control to improve energy efficiency and increase working time [22,23]. One model describing the energy consumption of a UAV is as follows [4,15]:

$$E = (\beta + \alpha h)t + P_{max}(h/s), \tag{3.5}$$

where E is the energy consumption during the period of t, β is the minimum power needed to hover just over the ground, α is a motor speed multiplier, s is the lifting speed, and $P_{max}(h/s)$ is the energy spent for lifting the robot to altitude h with speed s. A similar model has been discussed in [12]. Superior to [4,15], the model for recharging UAVs is also presented in [12], but the idea is similar. Another model characterizing the energy consumption for flying is given in [24] and such a model simply assumes that the energy consumption for flying is proportional to the flying distance. Readers are referred to [23], where a more complex energy consumption model is discussed.

3.2.4 Coverage optimization

With the models of coverage, connectivity, and energy consumption, in this subsection, some typical optimization problems are discussed.

3.2.4.1 Maximizing the number of covered targets

For the case with a given number of UAVs, if they are not sufficient to cover all the targets, one common objective is to maximize the number of covered targets. In general, the problems of maximizing the number of covered targets can be classified into two categories based on the input information about the users: exact location and general distribution.

Assuming the availability of targets' locations, a few publications provide some formulations for the deployment of a single UAV [25,26]. Specifically, an energy-efficient deployment approach for covering the maximum number of targets is presented in [25]. It first sets the vertical position of the UAV at the altitude providing the maximal coverage and then optimizes the horizontal position to maximize the number of covered targets while using minimum transmitting power. Instead of decoupling the deployment, reference [26] formulates a mixed-integer nonlinear problem involving both the horizontal coordinates and the altitude as variables, to maximize the number of served targets. Under the assumption of knowing targets' locations, the case of multiple UAVs has also been studied [27]. UAVs are used to serve the maximum number of targets, subject to that each robot has a capacity of service. The authors propose a K-means clustering-based algorithm. The reference [28] considers a UAV deployment problem with the objective of maximizing target coverage and minimizing the moving distance of UAVs when they need to change positions. A heuristic algorithm and a linear programming-based method are proposed.

For the case of multiple UAVs, another type of input information is the general distribution of targets [8,19,29]. It is worth mentioning that such a distribution differs from the long-term traffic behavior, which is usually used for the deployment of SBSs. The distribution considered here is for occasional events, such as sports games and concerts. The concept of target density is usually used to describe the distribution. Let $\rho(p)$ denote the target density of the location $p \in S$, where S is the area having targets to be covered by UAVs. Following the communication coverage model I, the total coverage area of the set of UAVs can be represented by $\mathcal{C}(P_1, \ldots, P_n) \in S$. Therefore, the maximization of the number of covered targets can be formulated as follows:

$$\max_{P_1, \ldots, P_n} \int_{p \in \mathcal{C}(P_1, \ldots, P_n)} \rho(p)dp. \tag{3.6}$$

In particular, reference [19] considers two objectives: the maximum covered number of targets and the energy consumption by UAVs for data transmission, together with the connectivity of UAVs with SBSs. Reference [8] follows the framework of [19] to address another optimization problem. The first goal is the maximization of the number of covered targets. The authors also consider the recharging of UAVs as well as interference management. Since this reference is based on the assumption of a street graph, the result is somewhat restrictive in practice.

3.2.4.2   Minimizing robot–user distance

Another commonly considered optimization problem is to minimize the average UAV–user distance [18]. To formulate this problem, the communication coverage model II is often used. Let $D_{min}(p)$ denote the distance from point $p \in S$ to the nearest UAV, which can be computed by [18]

$$D_{min}(p) = \min_{i=1,...,n} D(p, P_i). \tag{3.7}$$

Different from the disk model used in [19], the association of a user with a UAV in [18] is based on distance. Then, the objective to minimize the average UAV–user distance can be formulated as follows:

$$\min_{P_1,...,P_n} \int_{p \in S} \rho(p) D_{min}^2(p) dp, \tag{3.8}$$

where $\rho(p)$ is the user density at position $p \in S$ as mentioned above.

There are also some other publications based on user locations. Assuming the availability of user coordinates, the center of these users can be computed by taking the mean value of the coordinates [30]. Then, a UAV equipped with a tracking controller can track this center. Obviously, the center of users achieves the minimum average distance between users and the robot. Reference [31] aims at boosting the network capacity using UAVs. The authors propose a game-theory-based navigation algorithm. This algorithm is decentralized, but the computing load may be high since it uses exhaustive search to find a potential moving direction. Both these methods are based on the assumption of knowing the coordinates of all the users. However, how to obtain such information is not answered in them. Almost all the location-based approaches have not addressed such an issue either. In practice, measuring the coordinates is quite difficult or at least costly if some specific protocols are used. One solution is to estimate the locations of users. During the communication process with users, the UAVs can measure the received signal strength [32,33], from which the robots can estimate the locations of users in a dynamic manner using the robust extended Kalman filter [34]. Based on the estimated locations, a decentralized reactive navigation algorithm is presented in [35]. Different from [30], the UAVs in [35] move towards the weighted centers of mass. In other words, each user is assigned a dynamic weight and such weight depends on the distance between itself and the nearest robot. This is practical in real applications since a user which is closer to a robot will generally receive better quality of service than a user further away from the robot. Although [35] does not assume knowing the user locations, which is superior to [30] and many other location-based approaches, the performance of [35] depends on the accuracy of the location estimation. Another reactive approach is based on virtual forces [36]. The authors assume that the UAVs are able to recognize users by onboard sensors when they fly. They consider four types of virtual forces: hotspot attractive force, user attractive force, nearby robot repulsive force, and obstacle repulsive force. One weakness of [36] is that the locations of hotspots should always be detected before deploying the UAVs.

3.2.4.3 Minimizing the number of UAVs

The above approaches are about the deployment of a given number of UAVs. A different objective is to figure out the minimum number of required UAVs to provide a high quality of communication service or target coverage, which relates to the system investment. In [37], the authors deploy the UAVs at the same altitude to cover a set of targets, given their locations. Based on the disk coverage model, they formulate a geometric disk cover problem with the objective of using the minimum number of robots to cover all the targets. To address the problem, a centralized heuristic algorithm is proposed. Beyond the 2D situation considered in [37], the authors of [38] consider the case of deploying UAVs in 3D space with the same objective and a particle swarm optimization (PSO)-based heuristic algorithm is proposed. Reference [39] studies a similar problem and an elitist nondominated sorting genetic algorithm is used to find the optimal positions for UAVs from a given set of candidates.

Besides the application of supporting wireless communication using UAVs, the minimization of UAV number has also been considered in other applications such as wireless charging sensor nodes [15] and target tracking [4,24,40]. The publications [4,40] study the continuous camera coverage problem. The objectives are to minimize the number of UAVs and energy consumption. Reference [4] formulates a mixed-integer nonlinear optimization model and presents a mixed-integer programming-based heuristic algorithm. The paper [40] considers a similar case as [4] and the authors further propose a localized heuristic algorithm beyond the centralized one in [4]. It is worth mentioning that the advantages of [4,40] include the consideration of the energy limitation of UAVs. The paper [24] integrates the recharging requirements into the continuous coverage problem and examines the minimum number of UAVs for covering multiple subareas. The authors partition the coverage graph into cycles that start from the charging depot and the number of UAVs required depends on the charging time, the traveling time, and the number of subareas to be covered by the cycle.

3.2.4.4 Minimizing energy consumption

Since many UAVs are powered by onboard batteries, minimizing energy consumption improves operating lifetime. Also, when the serving object of UAVs is a WSN, the energy consumption of sensor nodes is often considered as a key point to ensure a long enough network lifetime.

Reference [41] considers the energy-efficient deployment of a UAV to provide wireless communication to targets. The received power at a target from a UAV is modeled as a function of the horizontal distance between the target and the UAV and the altitude of the UAV [41]. Then, the paper [41] decouples the optimal position in the horizontal and vertical dimensions. The horizontal position is taken as the center of the smallest circle covering the given set of targets, and then the vertical position is obtained such that the minimum transmitting power is used to cover that circle. The publication [12] focuses on the application of area monitoring. With a given set of UAVs and a charging depot on the ground, the authors consider the problem of how to schedule the UAVs to execute either monitoring task or recharging task, such that the network lifetime is maximized. They follow the camera coverage model and the energy consumption model therein is similar to Eq. (3.5). Another recent publication using a single UAV prolonged the network lifetime of a WSN [42]. Instead of minimizing the energy consumption of the UAV, the main points here are scheduling the wake-up time of sensor nodes and planning the trajectory of the UAV, such that the maximum energy consumption of sensor nodes is minimized.

3.2.4.5 Other optimization problems

Wirelessly charging sensor nodes in WSNs is a promising way to improve the lifetime of WSNs. The recent survey [43] reviews the related approaches using mobile ground robots as wireless chargers. In this subsection, some typical approaches using UAVs for wireless charging sensor nodes are reviewed.

Reference [44] presents some fundamental results related to charging sensor nodes using UAVs. A simple scenario with a single UAV is used, which can charge a single sensor at any time. With the objective of maximizing network lifetime, the authors aim at selecting appropriate nodes to recharge as well as the best sink node selection for data collection. Beyond this, the case where a UAV can recharge multiple sensor nodes has also been considered. Reference [45] considers the scenario of using a set of UAVs to collect data from a set of sensors and recharge them simultaneously. An optimization problem with the objective of maximizing data collection utility is proposed. A one-side-matching algorithm and a greedy algorithm are proposed to address the problem. The publication [46] moves beyond [45] by considering the varying energy consumption rates of sensors. A dynamic charging strategy is presented for a single UAV with the objective of minimizing the overall packet loss rate. The paper [47] focuses on the joint scheduling of charging routes and sensor association for multiple UAVs. The authors formulate a bounded route association problem and an approximation algorithm is proposed. The authors of [15] formulate a set cover problem to figure out the 3D positions of UAVs, with the objective of minimizing the UAV number such that all the sensor nodes can operate without time limitation.

3.2.5 Summary

In this section, some typical approaches for UAVs in various applications are discussed. To make a brief summary, a thorough comparison of these approaches by several common metrics is made, including the problem dimension, the number of UAVs, the UAV deployment manner, and the type of input information that the system uses for positioning the UAVs. In particular, the dimension of a problem can be 1D, 2D, and 3D, which depends on the application. The number of UAVs can be single or multiple. The UAV deployment manner can be divided into proactive and reactive. Generally, proactive deployment refers to the offline approaches, which compute the positions for the UAVs in advance; by contrast, reactive deployment refers to the online approaches, which can dynamically calculate robots' positions. For the type of information the system uses, two groups can be considered: location and density. The discussed approaches are summarized in Table 3.1. Table 3.1 presents the features of the discussed approaches under the aforementioned metrics and briefly summarizes the objectives of the corresponding optimization problems. Such a table may assist designers in picking the most useful approach according to the system features.

3.3 UAVs collaborating with WSNs

Different from the approaches discussed in Section 3.2, where UAVs are used as mobile sinks (mobile chargers) to collect sensory data (deliver energy) from (to) sensor nodes, this section discusses another application where WSNs provide service to UAVs, such as localization and navigation.

3.3.1 Localization of UAVs

A typical example is that UAVs are used to monitor key plants in the indoor factory environment, such as nuclear power stations, which is safer, more efficient, and more economical than human patrol. There are several challenges to be addressed about the navigation of UAVs. In indoor environments, GPS may not work, which makes localizing UAVs a challenge, as well as navigation. Although a camera can be equipped on the UAVs, it will lead to a heavy computing load because of image processing, which may be impossible for micro-UAVs with low computing performance [48]. To this end, WSNs can be used to localize UAVs in indoor environments using the extended Kalman filter and time difference of arrival (TDOA) measurements of radio signals [49].

TABLE 3.1 Summary of typical approaches for UAVs serving humans and WSNs.

Reference	UAV number	Dimension	Input	Remark
[14]	Single	1D	Location	Altitude optimization
[41]	Single	3D	Location	Min. transmit power
[25]	Single	3D	Location	Min. transmit power
[26]	Single	3D	Location	Max. covered number
[30]	Single	2D	Location	Tracking the center of users
[42]	Single	2D	Location	Prolong network lifetime
[37]	Multiple	2D	Location	Min. robot number
[19]	Multiple	2D	Density	Max. covered number
[8]	Multiple	2D	Density	Interference management
[24]	Multiple	2D	Location	Recharge sensors
[38]	Multiple	3D	Location	Min. UAV number
[39]	Multiple	3D	Location	Min. UAV number, max. data rate
[29]	Multiple	2D	Density	Neural-based cost function
[18]	Multiple	2D	Density	Decentralized robot–user distance minimization; connectivity
[27]	Multiple	2D	Location	K-means clustering
[20]	Multiple	3D	Location	Min. UAV number; connectivity
[4]	Multiple	3D	Location	Min. UAV number; energy constrained
[31]	Multiple	2D	Location	Exhaustive search
[35]	Multiple	2D	Distance	Move to weighted centers
[36]	Multiple	2D	Location	Navigation based on virtual force
[44]	Single	2D	Location	Selection charging node and sink node
[45]	Multiple	2D	Location	Max. data collection utility
[46]	Single	2D	Location	Varying energy consumption rates
[47]	Multiple	2D	Location	Charging routes and sensor association
[15]	Multiple	3D	Location	Min. UAV number

3.3.2 Navigation of UAVs

In WSNs, replacing a failed sensor node is a typical operation. To achieve this by UAVs, the capability to navigate a UAV towards a sensor node is required. One approach is based on received signal strength indication (RSSI) [50,51]. Specifically, in reference [50], the target sensor node periodically sends out beacons, and the UAV can measure RSSI to determine the moving direction. Under a similar setting, reference [51] presents a reduced particle filtering method, which is well suited for devices with limited computational power and energy resources.

The authors of [52] propose another system for navigating UAVs. Here, the destination of a UAV may not be a specific sensor node, but some other places in the industrial environment. The system consists of (1) a set of sensor nodes, equipped with 3D range-finder sensors, to detect the dynamic obstacles such as vehicles and walking people, (2) micro-UAVs, equipped with tracking controller only, and (3) a central controller. The micro-UAVs should measure their positions and directions and send such information to the central controller. The sensor nodes should also send their measuring data to the central controller. Based on the two types of information, a safe path is generated for each UAV by the central controller. Then, the UAVs, equipped with a simple tracking controller, track those paths to their destinations.

A similar publication to [52] is [53], where a team of UAVs perform surveillance without possessing sensors with automated target recognition capability and thus rely on communicating with unattended ground sensors placed on roads to detect and image potential intruders. The authors focus on the path planning problem with the objective of maximizing the likelihood of a UAV and an intruder being at the same location.

3.3.3 Summary

To briefly sum up, this section has discussed some basic tasks when UAVs collaborate with ground robots and WSNs. The former focuses on the application of search and rescue. The operations of target searching, path planning, and navigation are discussed. The latter mainly considers the localization and navigation of UAVs using a WSN. The discussed approaches are summarized in Table 3.2.

3.4 Discussion and future research directions

This section presents a discussion of the aforementioned approaches. At the same time, the discussion can raise some promising future research directions.

3.4.1 Connectivity consideration

As mentioned above, when multiple robots are used, each robot needs to have a valid wireless backhaul link at any time, to guarantee the delay of response, or form a connected backbone to transmit the collected sensory data to SBSs. References [18–20] have addressed this issue by requiring each robot to be connected to an SBS either directly or via another for relay, but the issues related to the data flow have not been covered in them. Under this model, the system will work as long as the connectivity is set up. However, the data rates at the one-hop robots are different from those at two-hop robots when serving users [9]. Therefore, such connectivity is not guaranteed to provide satisfactory service. One possible solution is inspired by the uneven clustering problem in WSNs

TABLE 3.2 Summary of typical approaches for UAVs collaborating with ground robots and WSNs.

Reference	Task	Collaboration type
[54]	Target searching using a ground robot	UAVs with ground robots
[55–57]	Target searching using a ground robot team	UAVs with ground robots
[58–60]	Target searching using UAVs	UAVs with UAVs
[61–66]	Reactive navigation for ground robots	UAVs with ground robots
[67,68]	Flying-ground robotic search and rescue team	UAVs with ground robots
[69–71]	Field inspection and parcel delivery with charging stations	UAVs with ground robots
[49]	Localization of UAVs by a WSN	UAVs with WSN
[50,51]	UAV navigation based on RSSI in WSNs	UAVs with WSN
[52,53]	UAV navigation by sensory information	UAVs with WSN

[72,73], i.e., setting different serving numbers for the UAVs with different numbers of hops, i.e., an unequal association between users and robots. In this way, the one-hop robots can serve a smaller number of users than the robots with two or more hops. Then, the one-hop robots can have more resources to relay the requests from the robots which are connected to them.

Another drawback of [18,19] is that they all fix the topology at first and then find optimal positions for UAVs satisfying this topology. It is clear that the topology of robots can also be optimized to achieve no worse performance of coverage. References [9,20] consider finding a subset of positions for the UAVs from a set of candidates. It is easy to understand that this method can generate different topologies of connected graphs, which is superior to [18,19].

3.4.2 Optimal deployment in 3D space

Among the discussed approaches, some are based on grids [4,40]. Although discretization simplifies the problem, the performance of the solution depends highly on the resolution of grids. The higher resolution makes the solution closer to the optimal one, but it increases the computing load, while the lower resolution makes the searching computational efficient, but the solution may be far from the optimal. Furthermore, there are some approaches based on the formulation of mixed-integer programming, such as [9,11,20,74]. One common feature of them is the assumption that the possible positions of UAVs are given by a set of candidates. These candidates can be regarded as a special set of grids.

From this discussion, it is believed that some efficiently computational algorithm should be developed for a case with continuous 3D deployment space.

3.4.3 Reactive deployment of UAVs

It can be seen from Table 3.1 that most available approaches for wireless communication support, target monitoring, etc., are proactive, while only a few are reactive. The reactive approaches are more suitable for dynamic situations. Thus, much more research efforts should be made on the development of reactive deployment methods. Furthermore, many of the existing publications assume the availability of targets' exact locations. It is clear that they are difficult to collect in practice, which impedes the applications of these methods. Therefore, a promising research direction is to develop location-free deployment algorithms.

3.4.4 Navigation with collision avoidance

In the application of search and rescue, navigation with collision avoidance is important for both ground and UAVs. Approaches such as [61,63,64,75] and some others such as [76–79] have proposed various path planning and reactive navigation algorithms. Although some of them are designed for UAVs, such as [76–78], where a fixed altitude is assumed, few papers studied the much more difficult case of collision-free 3D navigation [80–85]. With this regard, there is a necessity to extend the available 2D methods to the 3D scenario. Furthermore, 3D risk-aware navigation is preferred to avoid no-flight areas, such as populated areas and those with chemical plants [76–78]. Risk-aware navigation especially suits urban environments, where a relatively larger number of obstacles and no-flight areas exist. Another important direction for future research is to obtain some 3D versions of barrier and sweep coverage problems [86,87] for monitoring and surveillance applications.

3.4.5 Charging UAVs

Though many papers have presented strategies for UAVs in various applications, only a few of them have considered the battery recharging issue. Typically, small UAVs have fuel constraints, which prevent them from being used for long-term or large-scale missions.

Similar to the idea of using a ground robot as a mobile charging platform [69–71], an interesting research direction is to involve public transportation vehicles such as buses into the system. The UAVs can stay on top of these vehicles, travel together with them, and charge themselves simultaneously. Then, they can take off at some positions and head to their targets unreachable by these vehicles. This idea may simplify the flying-ground robotic system since the task is to plan paths for UAVs only. However, to make such a system operate efficiently, the timetables and the uncertainties in the timetables, e.g., due to traffic congestion in peak hours, should be accounted for in the planning stage.

3.4.6 RIS-assisted UAV communication

A new technology has attracted the attention of the wireless research community: reconfigurable intelligent surfaces (RISs). A RIS is comprised of a number of reconfigurable elements that can be controlled via integrated electronics [88]. RISs enable network operators to control the reflection, refraction, and scattering characteristics of the incident signals by inducing manageable phase shifts of the elements. Without complex decoding, encoding, and RF processing operations, RISs passively reflect the incident signals, and they are promising to effectively control the wireless communication environment. There are some publications investigating the optimization of phase shifts of RISs to enhance communication in several scenarios. A typical one is to enable a source-designation communication blocked by a building [89]. A more interesting scenario is to attach an RIS on a UAV. Then, the UAV can flexibly control the RIS so that certain ground users would be better served.

3.5 Summary

In this chapter, we presented an overview of the applications of UAVs in wireless communication systems. Specifically, the coverage issue, the connectivity issue, and the energy limitation constraint were discussed. The available publications for these issues were reviewed and comparisons were made to illustrate the features of them, through which shortcomings were pointed out. Some potential methods to address unsolved problems were indicated. Moreover, some promising research directions were discussed such as the optimal deployment of UAVs in 3D space, reactive deployment of UAVs, the charging issue of UAVs, and RIS-assisted UAV communication. These directions may open new topics that can benefit the research of UAVs.

References

[1] M. Erdelj, E. Natalizio, K.R. Chowdhury, I.F. Akyildiz, Help from the sky: leveraging UAVs for disaster management, IEEE Pervasive Computing 16 (1) (2017) 24–32.
[2] S.A.R. Naqvi, S.A. Hassan, H. Pervaiz, Q. Ni, Drone-aided communication as a key enabler for 5G and resilient public safety networks, IEEE Communications Magazine 56 (1) (2018) 36–42.
[3] C. Zhang, J.M. Kovacs, The application of small unmanned aerial systems for precision agriculture: a review, Precision Agriculture 13 (6) (2012) 693–712.
[4] L.D.P. Pugliese, F. Guerriero, D. Zorbas, T. Razafindralambo, Modelling the mobile target covering problem using flying drones, Optimization Letters 10 (5) (2016) 1021–1052.
[5] K. Kanistras, G. Martins, M.J. Rutherford, K.P. Valavanis, A survey of unmanned aerial vehicles (UAVs) for traffic monitoring, in: Handbook of Unmanned Aerial Vehicles, Springer, 2014, pp. 2643–2666.
[6] Z. Zhou, C. Zhang, C. Xu, F. Xiong, Y. Zhang, T. Umer, Energy-efficient industrial internet of UAVs for power line inspection in smart grid, IEEE Transactions on Industrial Informatics 14 (6) (2018) 2705–2714.

[7] B. Esakki, S. Ganesan, S. Mathiyazhagan, K. Ramasubramanian, B. Gnanasekaran, B. Son, S.W. Park, J.S. Choi, Design of amphibious vehicle for unmanned mission in water quality monitoring using internet of things, Sensors 18 (10) (2018).

[8] H. Huang, A.V. Savkin, A method for optimized deployment of unmanned aerial vehicles for maximum coverage and minimum interference in cellular networks, IEEE Transactions on Industrial Informatics (2018).

[9] C. Caillouet, T. Razafindralambo, Efficient deployment of connected unmanned aerial vehicles for optimal target coverage, in: Global Information Infrastructure and Networking Symposium (GIIS), IEEE, 2017, pp. 1–8.

[10] H. Huang, A.V. Savkin, Towards the internet of flying robots: a survey, Sensors 18 (11) (2018) 4038.

[11] D. Zorbas, T. Razafindralambo, F. Guerriero, et al., Energy efficient mobile target tracking using flying drones, Procedia Computer Science 19 (2013) 80–87.

[12] A. Trotta, M.D. Felice, F. Montori, K.R. Chowdhury, L. Bononi, Joint coverage, connectivity, and charging strategies for distributed UAV networks, IEEE Transactions on Robotics 34 (Aug 2018) 883–900.

[13] A. Al-Hourani, S. Kandeepan, A. Jamalipour, Modeling air-to-ground path loss for low altitude platforms in urban environments, in: Global Communications Conference (GLOBECOM), IEEE, Dec 2014, pp. 2898–2904.

[14] A. Al-Hourani, S. Kandeepan, S. Lardner, Optimal LAP altitude for maximum coverage, IEEE Wireless Communications Letters 3 (Dec 2014) 569–572.

[15] D. Zorbas, C. Douligeris, Computing optimal drone positions to wirelessly recharge IoT devices, in: INFOCOM Workshops, IEEE, 2018, pp. 628–633.

[16] M.Z. Win, P.C. Pinto, L.A. Shepp, A mathematical theory of network interference and its applications, Proceedings of the IEEE 97 (2) (2009) 205–230.

[17] Online: http://www.powercastco.com.

[18] A.V. Savkin, H. Huang, Deployment of unmanned aerial vehicle base stations for optimal quality of coverage, IEEE Wireless Communications Letters (2018).

[19] H. Huang, A.V. Savkin, An algorithm of efficient proactive placement of autonomous drones for maximum coverage in cellular networks, IEEE Wireless Communications Letters (2018).

[20] C. Caillouet, F. Giroire, T. Razafindralambo, Optimization of mobile sensor coverage with UAVs, in: INFOCOM Workshops, IEEE, 2018, pp. 622–627.

[21] E.P.D. Freitas, T. Heimfarth, I.F. Netto, C.E. Lino, C.E. Pereira, A.M. Ferreira, F.R. Wagner, T. Larsson, UAV relay network to support WSN connectivity, in: International Congress on Ultra Modern Telecommunications and Control Systems, Oct 2010, pp. 309–314.

[22] F. Jiang, A.L. Swindlehurst, Optimization of UAV heading for the ground-to-air uplink, IEEE Journal on Selected Areas in Communications 30 (June 2012) 993–1005.

[23] J. Xu, Y. Zeng, R. Zhang, UAV-enabled wireless power transfer: trajectory design and energy optimization, IEEE Transactions on Wireless Communications 17 (Aug 2018) 5092–5106.

[24] H. Shakhatreh, A. Khreishah, J. Chakareski, H.B. Salameh, I. Khalil, On the continuous coverage problem for a swarm of UAVs, in: The 37th Sarnoff Symposium, IEEE, 2016, pp. 130–135.

[25] M. Alzenad, A. El-Keyi, F. Lagum, H. Yanikomeroglu, 3D placement of an unmanned aerial vehicle base station (UAV-BS) for energy-efficient maximal coverage, IEEE Wireless Communications Letters 6 (4) (Aug. 2017) 434–437.

[26] I. Bor-Yaliniz, A. El-Keyi, H. Yanikomeroglu, Efficient 3-D placement of an aerial base station in next generation cellular networks, in: International Conference on Communications (ICC), Kuala Lumpur, Malaysia, IEEE, May 2016, pp. 1–5.

[27] B. Galkin, J. Kibilda, L.A. DaSilva, Deployment of UAV-mounted access points according to spatial user locations in two-tier cellular networks, in: Wireless Days (WD), March 2016, pp. 1–6.

[28] H. Huang, A.V. Savkin, M. Ding, M.A. Kaafar, C. Huang, On the problem of flying robots deployment to improve cellular user experience, in: The 37th Chinese Control Conference (CCC), July 2018, pp. 6356–6359.

[29] V. Sharma, M. Bennis, R. Kumar, UAV-assisted heterogeneous networks for capacity enhancement, IEEE Communications Letters 20 (June 2016) 1207–1210.

[30] Z. Becvar, M. Vondra, P. Mach, J. Plachy, D. Gesbert, Performance of mobile networks with UAVs: can flying base stations substitute ultra-dense small cells?, in: The 23th European Wireless Conference, VDE, 2017, pp. 1–7.

[31] A. Fotouhi, M. Ding, M. Hassan, Flying drone base stations for macro hotspots, IEEE Access 6 (2018) 19530–19539.

[32] P.N. Pathirana, A.V. Savkin, S. Jha, Location estimation and trajectory prediction for cellular networks with mobile base stations, IEEE Transactions on Vehicular Technology 53 (Nov 2004) 1903–1913.

[33] P.N. Pathirana, N. Bulusu, A.V. Savkin, S. Jha, Node localization using mobile robots in delay-tolerant sensor networks, IEEE Transactions on Mobile Computing 4 (May 2005) 285–296.

[34] I.R. Petersen, A.V. Savkin, Robust Kalman Filtering for Signals and Systems with Large Uncertainties, Birkhäuser, Boston, 1999.

[35] H. Huang, A.V. Savkin, Reactive deployment of flying robot base station over disaster areas, in: International Conference on Robotics and Biomimetics (ROBIO), Dec 2018.

[36] H. Zhao, H. Wang, W. Wu, J. Wei, Deployment algorithms for UAV airborne networks towards on-demand coverage, IEEE Journal on Selected Areas in Communications (2018).

[37] J. Lyu, Y. Zeng, R. Zhang, T.J. Lim, Placement optimization of UAV-mounted mobile base stations, IEEE Communications Letters 21 (March 2017) 604–607.

[38] E. Kalantari, H. Yanikomeroglu, A. Yongacoglu, On the number and 3D placement of drone base stations in wireless cellular networks, in: IEEE Vehicular Technology Conference, Sept 2016, pp. 1–6.

[39] S. Sabino, A. Grilo, Topology control of unmanned aerial vehicle (UAV) mesh networks: a multi-objective evolutionary algorithm approach, in: The 4th ACM Workshop on Micro Aerial Vehicle Networks, Systems, and Applications, DroNet'18, New York, NY, USA, ACM, 2018, pp. 45–50.

[40] D. Zorbas, L.D.P. Pugliese, T. Razafindralambo, F. Guerriero, Optimal drone placement and cost-efficient target coverage, Journal of Network and Computer Applications 75 (2016) 16–31.

[41] L. Wang, B. Hu, S. Chen, Energy efficient placement of a drone base station for minimum required transmit power, IEEE Wireless Communications Letters (2018).

[42] C. Zhan, Y. Zeng, R. Zhang, Energy-efficient data collection in UAV enabled wireless sensor network, IEEE Wireless Communications Letters 7 (June 2018) 328–331.

[43] H. Huang, A.V. Savkin, M. Ding, C. Huang, Mobile robots in wireless sensor networks: a survey on tasks, Computer Networks 148 (2019) 1–19.

[44] J. Johnson, E. Basha, C. Detweiler, Charge selection algorithms for maximizing sensor network life with UAV-based limited wireless recharging, in: The 8th International Conference on Intelligent Sensors, Sensor Networks and Information Processing, April 2013, pp. 159–164.

[45] Y. Pang, Y. Zhang, Y. Gu, M. Pan, Z. Han, P. Li, Efficient data collection for wireless rechargeable sensor clusters in harsh terrains using UAVs, in: Global Communications Conference, Dec 2014, pp. 234–239.

[46] L. Li, Y. Xu, Z. Zhang, J. Yin, W. Chen, Z. Han, A prediction-based charging policy and interference mitigation approach in the wireless powered internet of things, IEEE Journal on Selected Areas in Communications (2018).

[47] T. Wu, P. Yang, H. Dai, P. Li, X. Rao, Near optimal bounded route association for drone-enabled rechargeable WSNs, Computer Networks 145 (2018) 107–117.

[48] X. Liu, Z. Chen, W. Chen, X. Xing, Multiple optical flow sensors aiding inertial systems for UAV navigation, in: The 11th International Conference on Control (CONTROL), IEEE, 2016, pp. 1–7.

[49] J.-L. Rullán-Lara, S. Salazar, R. Lozano, Real-time localization of an UAV using Kalman filter and a wireless sensor network, Journal of Intelligent & Robotic Systems 65 (Jan 2012) 283–293.

[50] F. Bohdanowicz, H. Frey, R. Funke, D. Mosen, F. Neumann, I. Stojmenović, RSSI-based localization of a wireless sensor node with a flying robot, in: The 30th Annual ACM Symposium on Applied Computing, ACM, 2015, pp. 708–715.

[51] J. Radak, L. Baulig, D. Bijak, C. Schowalter, H. Frey, Moving towards wireless sensors using RSSI measurements and particle filtering, in: The 14th ACM Symposium on Performance Evaluation of Wireless Ad Hoc, Sensor, & Ubiquitous Networks, ACM, 2017, pp. 33–40.

[52] H. Li, A.V. Savkin, Wireless sensor network based navigation of micro flying robots in the industrial internet of things, IEEE Transactions on Industrial Informatics 14 (Aug 2018) 3524–3533.

[53] J. Las Fargeas, P. Kabamba, A. Girard, Cooperative surveillance and pursuit using unmanned aerial vehicles and unattended ground sensors, Sensors 15 (1) (2015) 1365–1388.

[54] B. Tovar, S.M.L. Valle, R. Murrieta, Optimal navigation and object finding without geometric maps or localization, in: International Conference on Robotics and Automation (ICRA), vol. 1, Sept 2003, pp. 464–470.

[55] W. Burgard, M. Moors, C. Stachniss, F.E. Schneider, Coordinated multi-robot exploration, IEEE Transactions on Robotics 21 (3) (2005) 376–386.

[56] M. Anderson, N. Papanikolopoulos, Implicit cooperation strategies for multi-robot search of unknown areas, Journal of Intelligent & Robotic Systems 53 (Dec 2008) 381–397.

[57] I.-K. Ha, Y.-Z. Cho, A probabilistic target search algorithm based on hierarchical collaboration for improving rapidity of drones, Sensors 18 (8) (2018).

[58] P. Doherty, P. Rudol, A UAV search and rescue scenario with human body detection and geolocalization, in: Australasian Joint Conference on Artificial Intelligence, Springer, 2007, pp. 1–13.

[59] M.A. Goodrich, B.S. Morse, D. Gerhardt, J.L. Cooper, M. Quigley, J.A. Adams, C. Humphrey, Supporting wilderness search and rescue using a camera-equipped mini UAV, Journal of Field Robotics 25 (1–2) (2008) 89–110.

[60] B.S. Morse, C.H. Engh, M.A. Goodrich, UAV video coverage quality maps and prioritized indexing for wilderness search and rescue, in: The 5th International Conference on Human-Robot Interaction, IEEE Press, 2010, pp. 227–234.

[61] J.M. Toibero, F. Roberti, R. Carelli, Stable contour-following control of wheeled mobile robots, Robotica 27 (1) (2009) 1–12.

[62] A.V. Savkin, C. Wang, Seeking a path through the crowd: robot navigation in unknown dynamic environments with moving obstacles based on an integrated environment representation, Robotics and Autonomous Systems 62 (10) (2014) 1568–1580.

[63] A.S. Matveev, H. Teimoori, A.V. Savkin, A method for guidance and control of an autonomous vehicle in problems of border patrolling and obstacle avoidance, Automatica 47 (3) (2011) 515–524.

[64] A.S. Matveev, M.C. Hoy, A.V. Savkin, The problem of boundary following by a unicycle-like robot with rigidly mounted sensors, Robotics and Autonomous Systems 61 (3) (2013) 312–327.

[65] A.S. Matveev, C. Wang, A.V. Savkin, Real-time navigation of mobile robots in problems of border patrolling and avoiding collisions with moving and deforming obstacles, Robotics and Autonomous Systems 60 (6) (2012) 769–788.

[66] A.V. Savkin, C. Wang, A simple biologically inspired algorithm for collision-free navigation of a unicycle-like robot in dynamic environments with moving obstacles, Robotica 31 (6) (2013) 993–1001.

[67] C. Shen, Y. Zhang, Z. Li, F. Gao, S. Shen, Collaborative air-ground target searching in complex environments, in: International Symposium on Safety, Security and Rescue Robotics, Conference, 2017, p. 230.

[68] J. Delmerico, E. Mueggler, J. Nitsch, D. Scaramuzza, Active autonomous aerial exploration for ground robot path planning, IEEE Robotics and Automation Letters 2 (2) (2017) 664–671.

[69] P. Maini, P. Sujit, On cooperation between a fuel constrained UAV and a refueling UGV for large scale mapping applications, in: International Conference on Unmanned Aircraft Systems (ICUAS), IEEE, 2015, pp. 1370–1377.

[70] Z. Luo, Z. Liu, J. Shi, A two-echelon cooperated routing problem for a ground vehicle and its carried unmanned aerial vehicle, Sensors 17 (5) (2017) 1144.

[71] K. Yu, A.K. Budhiraja, P. Tokekar, Algorithms for routing of unmanned aerial vehicles with mobile recharging stations and for package delivery, preprint, arXiv:1704.00079, 2017.

[72] H. Huang, A.V. Savkin, An energy efficient approach for data collection in wireless sensor networks using public transportation vehicles, AEÜ. International Journal of Electronics and Communications 75 (2017) 108–118.

[73] H. Huang, A.V. Savkin, Data collection in nonuniformly deployed wireless sensor networks by public transportation vehicles, in: The 85th Vehicular Technology Conference (VTC Spring), June 2017, pp. 1–4.

[74] F. Lagum, I. Bor-Yaliniz, H. Yanikomeroglu, Strategic densification with UAV-BSs in cellular networks, IEEE Wireless Communications Letters 7 (June 2018) 384–387.

[75] H. Huang, A.V. Savkin, Viable path planning for data collection robots in a sensing field with obstacles, Computer Communications 111 (2017) 84–96.

[76] A.V. Savkin, H. Huang, Optimal aircraft planar navigation in static threat environments, IEEE Transactions on Aerospace and Electronic Systems 53 (5) (2017) 2413–2426.

[77] S. Primatesta, G. Guglieri, A. Rizzo, A risk-aware path planning strategy for UAVs in urban environments, Journal of Intelligent & Robotic Systems (2018) 1–15.

[78] C. Yin, Z. Xiao, X. Cao, X. Xi, P. Yang, D. Wu, Offline and online search: UAV multiobjective path planning under dynamic urban environment, IEEE Internet of Things Journal 5 (2) (2018) 546–558.

[79] A.V. Savkin, H. Huang, The problem of minimum risk path planning for flying robots in dangerous environments, in: The 35th Chinese Control Conference (CCC), July 2016, pp. 5404–5408.

[80] H.L.N.N. Thanh, N.N. Phi, S.K. Hong, Simple nonlinear control of quadcopter for collision avoidance based on geometric approach in static environment, International Journal of Advanced Robotic Systems 15 (2) (2018) 1729881418767575.

[81] H.L.N.N. Thanh, S.K. Hong, Completion of collision avoidance control algorithm for multicopters based on geometrical constraints, IEEE Access 6 (2018) 27111–27126.

[82] T. Elmokadem, A 3D reactive collision free navigation strategy for nonholonomic mobile robots, in: The 37th Chinese Control Conference (CCC), IEEE, 2018, pp. 4661–4666.

[83] X. Yang, L.M. Alvarez, T. Bruggemann, A 3D collision avoidance strategy for UAVs in a noncooperative environment, Journal of Intelligent & Robotic Systems 70 (1–4) (2013) 315–327.

[84] C. Wang, A.V. Savkin, M. Garratt, A strategy for safe 3D navigation of non-holonomic robots among moving obstacles, Robotica 36 (2) (2018) 275–297.

[85] A.V. Savkin, A.S. Matveev, M. Hoy, C. Wang, Safe Robot Navigation among Moving and Steady Obstacles, Elsevier, 2015.

[86] T.M. Cheng, A.V. Savkin, Decentralized control for mobile robotic sensor network selfdeployment: barrier and sweep coverage problems, Robotica 29 (2) (2011) 283–294.

[87] A.V. Savkin, T.M. Cheng, Z. Xi, F. Javed, A.S. Matveev, H. Nguyen, Decentralized Coverage Control Problems for Mobile Robotic Sensor and Actuator Networks, IEEE Press – Wiley, 2015.

[88] Q. Wu, R. Zhang, Towards smart and reconfigurable environment: intelligent reflecting surface aided wireless network, IEEE Communications Magazine 58 (1) (2019) 106–112.

[89] H.-M. Wang, J. Bai, L. Dong, Intelligent reflecting surfaces assisted secure transmission without eavesdropper's CSI, IEEE Signal Processing Letters 27 (2020) 1300–1304.

Chapter 4

Data collection in wireless sensor networks by ground robots with full freedom[☆]

4.1 Motivation

Mobile sinks are regarded as a promising method to prolong the lifetime of WSNs. By physical movement, mobile sinks can save much energy for sensor nodes, since the communication between them can be done over a relatively short distance. As a coin has two sides, one defect of such a scheme is that the data delivery delay is also increased due to the relatively slow physical movement. Thus, one important research topic is how to design the path for mobile sinks such that all the sensory data can be collected and at the same time the delivery delay is minimized. One basic approach is to view such problem as the traditional TSP [1]. The mobile sink is regarded as a salesman and the sensor nodes are regarded as the delivery destinations. Then, the problem is to find the shortest path in length such that every sensor node is visited exactly once. When multiple mobile sinks are used, the corresponding problem can be viewed as the VRP [2].

Unlike the pointwise model in TSP- and VRP-based approaches, in this chapter, the mobile sinks are modeled as a unicycle moving at a constant speed with bounded angular speed. This model is also called the Dubins car [3], and it is well known that the motion of many wheeled robots and UAVs can be described by this model [4–7]. Recent studies have proposed various solutions to the path planning problem for data collection by mobile robots. However, several issues still need further investigation. One is that many existing approaches set the sensor nodes' locations as target positions for the robots (e.g., see [8]), which leads to collisions with sensor nodes. Another issue is that most studies assume the sensing field is obstacle-free (e.g., see [9]), which is quite ideal in practice. Furthermore, the dynamic constraint of robots is rarely taken into account in recent algorithms. Then, the produced paths may have nonsmooth corners [10]. This is a severe limitation in practice because the paths cannot be applied to some robots such as the considered unicycle robots with the

---

☆ The main results of the chapter were originally published in Hailong Huang, Andrey V. Savkin, Viable path planning for data collection robots in a sensing field with obstacles, Computer Communications 111 (2017) 84–96. Permission from Elsevier for reuse was obtained.

Copyright © 2022 Elsevier Inc. All rights reserved.

bounded angular speed. Some approaches assume the data transfer time from the source sensor nodes to the robots is negligible [11]. It is reasonable for the light data load nodes, but not for the nodes with heavy data load, such as the nodes equipped with cameras to snapshot speeding vehicles [12].

With the mentioned concerns, we define a *viable* path, which is smooth, collision-free with sensor nodes/base station and obstacles, and closed and provides enough contact time with all the sensor nodes. The viable path takes into account the properties of both robots and sensor networks, which is close to reality. The main objective is to design the shortest viable path for the considered robots. We formulate the problem as a variant of the DTSPN [13,14]. To solve the problem, we propose an SVPP algorithm. In essence, this algorithm is based on a tangent graph, which is then modified by a reading adjustment to provide enough contact time for each node. With the modified tangent graph, it determines a permutation of sensor nodes by solving an asymmetric TSP (ATSP) instance. Having the permutation, the modified tangent graph is simplified and converted to a tree-like graph, where the edges and vertices unrelated to the permutation are removed. Finally, the shortest viable path is obtained by searching the tree-like graph using a dynamic programming-based method.

The viable paths produced by the SVPP algorithm can be traveled by the considered unicycle robots periodically. However, such paths are designed for a single mobile sink. For the large-scale network, we further consider the situation where multiple mobile sinks are available, which is promising to reduce the average path length for each sink and shorten the delivery delay. We consider k identical mobile sinks and target on designing viable paths for them such that the lengths of the k paths are more or less equal. Then the data delivery delays on these paths can be similar. We first discuss an algorithm by introducing a viable path to k-SPLITOUR [15] (denoted as viable k-SPLITOUR), which constructs a whole path using SVPP and then splits it into k subpaths of more or less equal lengths. We point out that the generated k viable paths are not guaranteed to be optimal. With this regard, we conduct a further operation, i.e., we reconstruct the k paths using SVPP; this algorithm is referred to as k-SVPP. It is easy to understand that k-SVPP achieves no worse performance than viable k-SPLITOUR, and in many cases, k-SVPP performs better.

For performance evaluation, extensive simulations are conducted. First, SVPP is applied to obtain viable paths for the network instances with different numbers of nodes and topologies where obstacles exist. We investigate the influences of two factors, robot speed and data load, on the system metrics path length and collection time. Second, the performance of k-SVPP is demonstrated by comparing with viable k-SPLITOUR. Third, the comparison of the proposed methods with a multihop communication algorithm (shortest path routing) is also provided. We study the influence of data load distributions on data collection performance in terms of energy consumption. We find that using mobile robots to collect data saves around 95% energy for sensor nodes compared to multihop communication, which increases the network lifetime. The main results of the chapter were originally published in [16]. The key contributions of this chapter include the following:

1. We propose the concept of viable path, which combines the concerns of robotics and sensor networks.
2. We provide a formulation for the studied problem as a variant of DTSPN and an algorithm called SVPP.
3. We provide a k-SVPP algorithm to design k viable paths which have approximately equal lengths.
4. We present extensive simulations to demonstrate the effectiveness and advantages of our algorithms.
5. We further consider the problem of navigating a military aircraft in a threat environment to its final destination while minimizing the maximum threat level and the length of the aircraft path. The proposed method to construct optimal low-risk aircraft paths involves a simple geometric procedure and is very computationally efficient.

The remaining parts of this chapter are structured as follows. Section 4.2 formally describes the studied problem. Section 4.3 provides the suggested SVPP algorithm, and a k-SVPP algorithm dealing with path planning for k robots is proposed in Section 4.4. Section 4.5 demonstrates the performance of our algorithms by simulations and comparisons with the alternatives. A discussion is provided in Section 4.6. Finally, Section 4.7 concludes the chapter.

4.2 System model and problem statement

This section describes the system model, gives the basic assumptions, and states our problem formally. The main notations used in this chapter are listed in Table 4.1.

Consider a planar robot modeled as a unicycle, whose state can be represented as the configuration $X = (x, y, \theta) \in SE(2)$, where $(x, y) \in \mathbb{R}^2$ is the robot's position and $\theta \in \mathbb{S}^1$ is its heading. The robot travels with a constant speed v and is controlled by the angular speed u. The dynamics of the robot can be described as

$$
\begin{aligned}
\dot{x}(t) &= v \cos \theta(t), \\
\dot{y}(t) &= v \sin \theta(t), \\
\dot{\theta}(t) &= u(t) \in [-u_M, u_M],
\end{aligned}
\tag{4.1}
$$

where u_M is the given maximum angular speed. Such a model describes a planar motion of many ground robots, missiles, UAVs, underwater vehicles, etc. [4,7, 17,18]. The standard nonholonomic constraint can be represented as

$$
|u(t)| \le u_M.
\tag{4.2}
$$

TABLE 4.1 Notations.

Notation	Comment
X	Configuration of robot
(x, y)	Position of robot
θ	Heading of robot
u	Angular speed of robot
v	Speed of robot
u_M	Maximum angular speed of robot
R_{min}	Minimum turning radius of robot
n	The number of base stations and nodes
s_i	Base station or sensor node
m	The number of obstacles
o_j	Obstacle
d_{safe}	Obstacle safety margin
∂o_j	Boundary of the convex hull
C_i	Visiting circle
$G(V, E)$	(Modified) tangent graph
$\mathcal{L}$	Viable path construction function
g	Data load
r	Data transmission rate
e	Transmission energy consumption rate
δ	Required contact time
Σ	Permutation of visiting with length n
σ_i	The ith element of Σ
$G(V', E')$	Simplified tangent graph
$\mathcal{P}$	Projection function from $SE(2)$ to $\mathbb{R}^2$
K	The blocking number
Σ'	Extended permutation ($n' = n + K$)
σ_i'	The ith element of Σ'
$\mathcal{L}'$	Viable path construction function
T	Tree-like graph, where T_i is the ith layer of T
L	Path length
z	Configuration variable
P	Viable path
k	The number of robots
len	Path length function

Then the minimum turning radius of the robot is

$$R_{min} = \frac{v}{u_M}. \tag{4.3}$$

Any path $(x(t), y(t))$ of the robot (4.1) is a plane curve satisfying the following constraint on its so-called average curvature (see [3]). Let $P(a)$ be this path parametrized by arc length. Then

$$\|P'(a_1) - P'(a_2)\| \le \frac{1}{R_{min}}|a_1 - a_2|. \qquad (4.4)$$

Here $\| \cdot \|$ denotes the standard Euclidean vector norm. Note that we use the constraint (4.4) on average curvature because we cannot use the standard definition of curvature from differential geometry [19] since the curvature may not exist at some points of the robot path.

Consider a sensor network deployed in a cluttered environment. It consists of a base station (s_1) and $n - 1$ sensor nodes $(s_i, i \in [2, n])$. We overload s_i $(i \in [1, n])$ as a sensor node/base station and its location. Now s_1 executes the path planning task and the robot downloads the path before it departs. Consider m disjoint and smooth obstacles $(o_j, j \in [1, m])$ in the field. The locations of all the stationary sensor nodes and the obstacles as well as the shapes of the obstacles are known. Let d_{safe} be the safety margin for the obstacles. Then we can get the safety boundary of o_j (see Fig. 4.1b for an example). Since o_j can be a nonconvex set, its safety boundary can also be nonconvex. As discussed below, our objective is to pass the obstacles as fast as possible. Then, moving along the boundary of the convex hull of o_j saves time comparing to moving along the winding safety boundary. Let ∂o_j $(j \in [1, m])$ be the boundary of the convex hull of o_j's safety boundary. The way to construct such a convex hull is simple. For the nonconvex part, a common tangent line is placed (see Fig. 4.1b for an example). Regarding constraint (4.4), we assume that any ∂o_j $(j \in [1, m])$ is a smooth curve with the curvature $c(p)$ at any point p satisfying $c(p) \le \frac{1}{R_{min}}$.

The objective is to design the shortest viable path for the robot (4.1). We define the viable path as follows.

Definition 4.2.1. A path P is viable if it is smooth, collision-free, and closed, and offers enough contact time to read all the data from the sensor nodes.

The motivations behind the viable path include the following practical concerns. First, as mentioned, based on the considered robots' bounded turning radius, it is inappropriate for them to turn at a spot. Thus, the foremost condition of the viable path is smoothness. Second, the robot should not traverse the locations of the sensor nodes; otherwise, collisions may happen. Besides, since obstacles may exist in the sensing field, the robot should also be able to avoid them. Furthermore, as many applications require the robot to execute the data collection task periodically, the closeness feature may facilitate this requirement. Finally, to read all the sensory data from sensor nodes, the produced path should provide enough time for the robot to communicate with sensor nodes. Next, we detail how to design viable paths.

Concerning condition (2) and considering the dynamics (4.1), we first describe our visiting circle model.

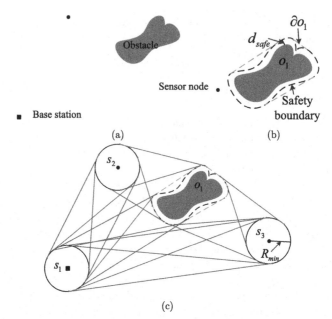

FIGURE 4.1 An illustrative example with one base station, two nodes, and one nonconvex obstacle. (a) Sensing field. (b) Convex hull construction. (c) Tangent graph.

Definition 4.2.2. The visiting circle is centered at the location of a sensor node/base station with radius R_{min}.

The radius of the visiting circle can be smaller than R_{min}. But as R_{min} is the minimum turning radius, the robot cannot move along a circle whose radius is smaller than R_{min}. Being able to move along the visiting circles facilitates our work in the applications where the stored data cannot be uploaded in a short time. In this case, the robot can rotate around the sensor nodes/base station on the visiting circles until the data uploading is finished. Thus, the radius of the visiting circle is set as R_{min}. Similar to d_{safe}, R_{min} can also be regarded as the safety margin of sensor nodes/base stations. We use C_i, $i \in [1, n]$, to represent the visiting circle and we use the term *element* to represent either a visiting circle C_i ($i \in [1, n]$) or a boundary ∂o_j ($j \in [1, m]$) in the rest of this chapter. We assume that any two elements do not intersect.

Now we construct the tangent graph [7]. One major component of the tangent graph is the *tangent*, defined as a straight line that is simultaneously tangent to two elements and not intersecting with any others. The common point of a tangent and an element is called the *tangent point*. The curve between two tangent points on the same element is called the *arc*. Let $G(V, E)$ be the tangent graph, where vertex set V consists of a finite set of tangent points and edge set E consists of a finite set of tangents and arcs. Fig. 4.1 depicts an example of $G(V, E)$.

We claim that the path for the robot can be formed by a subset of E, since when the robot moves on $G(V, E)$, conditions (2) and (3) are satisfied definitely. However, such a path is not viable. To make the path smooth, we further consider a heading constraint.

Definition 4.2.3. The heading constraint refers to that the robot's heading θ at the beginning of an edge should be equal to that at the ending of the last edge.

Let $edge_1, edge_2 \in E$ be two edges and let them have a common tangent point p. Let $\theta^p_{edge_1}$ and $\theta^p_{edge_2}$ be the robot's headings at the ending of edge $edge_1$ and the beginning of edge $edge_2$, respectively, i.e., at point p. Then, the heading constraint requires

$$\theta^p_{edge_1} = \theta^p_{edge_2}. \tag{4.5}$$

By carefully selecting the edges according to (4.5), the path can be smooth, i.e., condition (1) is satisfied.

Now we consider condition (5), which relates to the communication mode. We assume that the buffered data are transmitted only when the robot (4.1) is on C_i, $i \in [1, n]$. Note that as long as the robot is within the transmitting range of the sensor nodes, the sensor nodes can transmit data to the robot even before it arrives at the visiting circle and after it leaves. But it is practical to consider the aforementioned assumption. One reason is that short-distance communication guarantees a low data loss rate. Besides, it results in a relatively low transmission energy consumption. With respect to this assumption, the node's data load and data transmission rate together determine the required contact time. However, the arc edge constructed by two tangent points on a visiting circle does not guarantee such contact time. So we need to adjust the arc edge when necessary. Let g be the data load of a node over a period $\mathcal{T}$ and let r be the data transmission rate. We consider the nonincreasing staircase model [20] where the data transmission rate $r(d)$ (bit/s) is a function of distance (d) between itself and the robot:

$$r(d) = \begin{cases} r_1, 0 < d \leq d_1, \\ r_2, d_1 < d < d_2, \\ \vdots \\ r_i, d_{i-1} < d \leq d_i, \\ \vdots \\ 0, d > d_{max}, \end{cases} \tag{4.6}$$

where d_{max} is the transmission range of a sensor node. Further, each data transmission rate is associated with an energy consumption rate $e(d)$ (J/bit), which is modeled as a nondecreasing staircase function of d.

For s_i, the required contact time δ_i can be calculated by

$$\delta_i = \frac{g_i}{r(R_{min})}. \tag{4.7}$$

With speed v, we obtain the minimum arc length l_i:

$$l_i = \delta_i v = \frac{g_i v}{r(R_{min})}. \tag{4.8}$$

We introduce two parameters to describe our arc edge adjustment model:

$$\begin{aligned} a &= \left\lfloor \frac{g_i v}{r(R_{min}) 2\pi R_{min}} \right\rfloor, \\ b &= \frac{g_i v}{r(R_{min})} - a \cdot 2\pi R_{min}. \end{aligned} \tag{4.9}$$

Given a visiting circle C_i, $i \in [1, n]$, and all the arcs on C_i, any arc $\widehat{p_1 p_2}$ can be adjusted by

$$|\widehat{p_1 p_2}| = \begin{cases} a \cdot 2\pi R_{min} + |\widehat{p_1 p_2}|, & \text{if } (i), \\ (a+1) \cdot 2\pi R_{min} + |\widehat{p_1 p_2}|, & \text{if } (ii), \end{cases} \tag{4.10}$$

where

$$\begin{aligned} (i) &: b \le |\widehat{p_1 p_2}|, \\ (ii) &: b > |\widehat{p_1 p_2}|. \end{aligned}$$

With the adjustment (4.10), $\widehat{p_1 p_2}$ is a valid arc edge allowing all the buffered data to be read by the robot. So we call (4.10) the reading adjustment. If $\widehat{p_1 p_2}$ is selected as a part of path, it means the robot arrives at p_1, makes a or $a+1$ more round trips around s_i, and leaves from p_2. In the rest of this chapter, $G(V, E)$ represents the modified tangent graph, where V remains the same and the arc edges on visiting circles in E are adjusted by (4.10). With this model, extracting a subset of E satisfies condition (5).

Finally, with respect to condition (4) and the shortest path request, we introduce the problem to be addressed as the DTSPN in [14,21]. Let $C = \{C_1, ..., C_n\}$ be a set of goal regions to be visited and let $\Sigma = \{\sigma_1, ..., \sigma_n\}$ be an ordered permutation of $\{C_1, ..., C_n\}$. Define a projection from $SE(2)$ to $\mathbb{R}^2$ as $\mathcal{P}: SE(2) \to \mathbb{R}^2$, i.e., $\mathcal{P}(X) = (x, y)$. The considered problem is an optimization over all possible permutations Σ and configurations X. Stated more formally,

$$\min_{\Sigma, X} |\mathcal{L}(X_{\sigma_n}, X_{\sigma_1})| + \sum_{i=1}^{n-1} |\mathcal{L}(X_{\sigma_i}, X_{\sigma_{i+1}})| \tag{4.11}$$

$$s.t. \ \mathcal{P}(X_{\sigma_i}) \in \sigma_i, \ i \in [1, n],$$

where $\mathcal{L}(X_1, X_2)$ is the viable subpath (defined below) with minimum length from configuration X_1 to X_2 ($\mathcal{P}(X_1) \in C_{i_1}$, $\mathcal{P}(X_2) \in C_{i_2}$, $i_1, i_2 \in [1, n]$, and $i_1 \neq i_2$) and $|\mathcal{L}(X_1, X_2)|$ gives the length.

Definition 4.2.4. A viable subpath consists of a subset of E. Any two consecutive edges on the viable subpath have a common tangent point where the heading constraint (4.5) is satisfied.

So far, since the function $\mathcal{L}$ makes the generated paths satisfy conditions (1), (2), (3), and (5) and the formulation of the problem (4.11) considers condition (4) as well as the shortest request, the path produced by solving the problem (4.11) is the shortest viable path for the data collection unicycle robot. The next section discusses how to address this problem.

4.3 Shortest viable path planning

Formulated as the DTSPN like [13,14], problem (4.11) is also NP-hard. But unlike the DTSPN, having an infinite number of possible configurations in each goal region, the candidate configuration number on C_i, $i \in [1, n]$, in problem (4.11) is finite due to the tangent graph. So more appropriately, problem (4.11) is a sampled DTSPN. On the other hand, in the context of sensor networks, sensor nodes are usually deployed apart from each other. Then our considered problem is a standard version of the sampled DTSPN.

We propose an algorithm called SVPP. The basic idea is similar to [21], where the first stage is to determine the permutation Σ based on $G(V, E)$. In the second stage, we simplify $G(V, E)$ into $G(V', E')$ and convert $G(V', E')$ to a tree-like graph T, and then search the shortest path in T. The main steps of SVPP are outlined as Algorithm 1. The details are given below.

Algorithm 1 Shortest viable path planning (SVPP).

1: Compute Σ by solving the ATSP instance based on $G(V, E)$.
2: Compute Σ' by adding blocking safety boundaries to Σ.
3: Simplify $G(V, E)$ to $G(V', E')$ by retaining the edges and vertices related to Σ' and deleting the others.
4: Convert $G(V', E')$ to the tree-like graph T.
5: Given an initial configuration, search the shortest path P in T.

In Step 1, to compute Σ, we construct a directed graph. The reason for choosing a directed graph instead of an undirected graph lies in condition (5) of Definition 4.2.1, i.e., different sensor nodes may require different contacting times. We take s_i, $i \in [1, n]$, as the vertex set of the directed graph. We construct the edge set as follows. The length of the edge between two vertices takes into account two aspects: the length of the valid path between their visiting circles and the length of the adjusted arc on the latter vertex. Note that here, since there may be multiple valid paths between two visiting circles (see s_1 and s_2 in

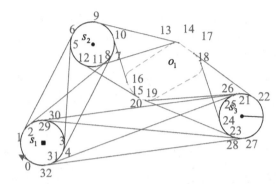

FIGURE 4.2 Simplified tangent graph $G(V', E')$ of $G(V, E)$ in Fig. 4.1c. Since $\Sigma = \{C_1, C_2, C_3\}$, the edges and vertices between C_1 and ∂o_1 are removed while the others are retained. Since ∂o_1 blocks C_2 and C_3, ∂o_1 is inserted into Σ. Then we obtain $\Sigma' = \{C_1, C_2, \partial o_1, C_3\}$. The numbers near the vertices are the labels of robot configurations. For example, label 0 represents X_0.

Fig. 4.2 for an example), we use the average length of them. Thus, the length of the edge from s_1 to s_2 equals the summation of the average length and the length of the adjusted arc on the visiting circle of s_2. In contrast, the length of the edge from s_2 to s_1 equals the summation of the average length and the length of the adjusted arc on the visiting circle of s_1. With such a directed graph, we use an ATSP solver [22] to calculate the permutation Σ.

Having Σ, $G(V, E)$ can be simplified by retaining the tangent edges connecting two successful visiting circles in Σ and the corresponding arc edges. When any obstacle blocks any pair of visiting circles, the edges passing the obstacle safety boundaries are also retained. By doing this, we can get the blocking number $K \geq 0$ which counts the occurrences of blocking two successful visiting circles in Σ. Since one obstacle can block more than one pair of successful visiting circles, the number of blocking obstacles in $G'(V', E')$ may be smaller than K. We insert these boundaries into the proper positions in Σ and get an extended permutation $\Sigma' = \{\sigma'_1, ..., \sigma'_{n'}\}$, whose length is $n' = n + K$. Obviously, $\Sigma \subseteq \Sigma'$. We call the new graph the simplified tangent graph $G'(V', E')$, where $V' \subseteq V$ and $E' \subseteq E$. Given Σ', the problem (4.11) can be reformulated as

$$\min_X |\mathcal{L}'(X_{\sigma'_{n'}}, X_{\sigma'_1})| + \sum_{i=1}^{n'-1} |\mathcal{L}'(X_{\sigma'_i}, X_{\sigma'_{i+1}})| \qquad (4.12)$$

$$s.t. \ \mathcal{P}(X_{\sigma'_i}) \in \sigma'_i, \ i \in [1, n'],$$

where $\mathcal{L}' = \mathcal{L}$.

It is worth mentioning that if we name X_1 (X_2) as the arrival configuration on C_{i_1} (C_{i_2}), there is a departure configuration, say X_3, on C_{i_1} such that the heading constraint (4.5) is satisfied. Then $\mathcal{L}'(X_1, X_2)$ always consists of two edges: one arc edge $\overparen{\mathcal{P}(X_1)\mathcal{P}(X_3)}$ and one tangent edge $\overline{\mathcal{P}(X_3)\mathcal{P}(X_2)}$. With such characteristic of $\mathcal{L}'$, the objective of problem (4.12) can be stated to find the minimum-length path by selecting two configurations from each element in $G(V', E')$, subject to the heading constraint (4.5).

Before we describe how to select such configurations, we analyze the number of possible paths in $G(V', E')$. We generate tangents between any pair of elements to construct $G(V, E)$. As the robots only need to move along the boundary of the convex hull, we have four common tangents and eight tangent points for each pair of successful elements in $G(V', E')$. Then one element has eight configurations, four of which are arrival configurations and the other four are departure configurations. For example, in Fig. 4.2, X_5, X_6, X_7, and X_8 are the four arrival configurations of C_2 and X_9, X_{10}, X_{11}, and X_{12} are the four departure configurations. Also, we note that from one arrival configuration of one element, we have two options to reach the arrival configurations of the next element considering the heading constraint (4.5). Then, from a given X_0 we have $2^{n'}$ paths to reach $\sigma'_{n'}$. Finally, the robot needs to return to σ'_1 from $\sigma'_{n'}$. Again due to the heading constraint (4.5), half of the arrival configurations on $\sigma'_{n'}$ cannot reach X_0, such as X_{26} and X_{28} in Fig. 4.2. Therefore, the total number of paths starting and ending at X_0 is $2^{n'-1}$.

To better demonstrate the viable paths in $G(V', E')$, we convert it to a tree-like graph. Given X_0, we cut σ'_1 into two parts: one part contains the departure configurations and X_0, and the other contains the arrival configurations and X_0. Then the circle-like graph $G(V', E')$ turns out to be a tree-like graph, which is called tree-like graph T. Since n' elements are on $G(V', E')$ and σ'_1 is divided into two parts, T consists of $n' + 1$ layers, where $T_i = \sigma'_i$ ($i \in [1, n']$) and $T_{n'+1} = \sigma'_1$. An example of T is shown in Fig. 4.3.

Evolved from $G(V', E')$, T inherits the feature that given a departure configuration, moving from one layer to the next one has only one path. Define z_i ($i \in [1, n' + 1]$) as an arrival configuration variable of layer T_i. Define $L_{i,i+1}$ as the path length from an arrival configuration z_i on T_i to an arrival configuration z_{i+1} on T_{i+1}. Then we have

$$L_{i,i+1} = |\mathcal{L}'(z_i, z_{i+1})|, \ 1 \le i \le n'. \tag{4.13}$$

Suppose $L^*_{1,i+1}$ is the shortest path length from an arrival configuration on T_1 to an arrival configuration on T_{i+1}, $1 \le i \le n'$. Since the initial value of z_1 is prescribed, i.e., $z_1 = X_0$, it follows that $L^*_{1,i+1}$ is only a function of the viable z_{i+1}. Then

$$L^*_{1,i+1} = \min \{L^*_{1,i} + L_{i,i+1}\}, \ 1 \le i \le n', \tag{4.14}$$

where $L^*_{1,1} = 0$.

Let z^*_i, $i \in [1, n' + 1]$, denote the optimal configurations to problem (4.14). We use a dynamic programming-based method, given by Algorithm 2, to solve problem (4.14).

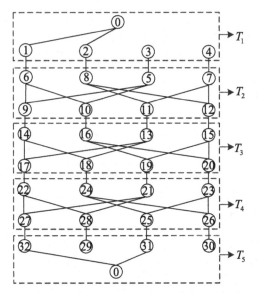

FIGURE 4.3 Tree-like graph of the example shown in Fig. 4.2. There are eight viable paths starting at X_0 and ending at X_0.

Algorithm 2 Shortest path search.

Require: $T, \mathcal{L}'$
Ensure: P
1: **for** $i = 2, i \leq n', i = i + 1$ **do**
2:     For each feasible $z_i \in T_i$, choose feasible $z_{i-1} \in T_{i-1}$ to minimize $L^*_{1,i} + L_{i,i+1}$. Then $L^*_{1,i+1}$ is expressed as a function of z_i only.
3: **end for**
4: Use the destination configuration z_1 to select configurations $z^*_{n'+1}, z^*_{n'},...,$ z^*_2. $z^*_1 = z_1$.
5: $P = \{\mathcal{L}'(z^*_1, z^*_2), ..., \mathcal{L}'(z^*_{n'}, z^*_{n'+1}), \widehat{\mathcal{P}(z^*_{n'+1})\mathcal{P}(z^*_1)}\}$.

To end this section, we analyze the time complexity of SVPP. In Step 1, the complexity of the ATSP algorithm [22] is $O(n^3)$. In Step 2, we check n pairs of visiting circles to see whether they are blocked by any boundary of the convex hull. In each checking, the worst case is to check all the m obstacles and then the time complexity is $O(mn)$. In Step 3, we perform a constant number of operations to each element in Σ'. Then the time complexity of the simplifying procedure is $O(n + K)$. Converting $G(V', E')$ to T costs $O(1)$ in Step 4. Finally, because the maximum number of arrival configurations is four on each layer of T, searching the shortest path in T costs $O(n + K)$. Therefore, the overall time complexity of SVPP is $O(n^3)$.

4.4 k-Shortest viable path planning

In Section 4.3, we have only considered the situation of using a single robot. Applying SVPP to a large-scale network may result in a path with an unacceptable collection time. One possible solution to deal with the long collection time is to employ multiple robots. Given the sensor network, increasing the number of robots can decrease the average path length of the robots. In this scenario, we need to design a viable path for each robot. Given k identical robots whose initial positions are at $\mathcal{P}(X_0) \in C_1$, we aim at finding k viable paths $\{P_1, ..., P_k\}$ such that:

1. $P_l, l \in [1, k]$, starts from $\mathcal{P}(X_0)$ and ends at C_1;
2. for each $i \in [2, n]$, $\exists l \in [1, k]$ such that (1) $C_i \cap P_l \neq \emptyset$ and (2) $C_i \cap P_j = \emptyset$, $j \in [1, k]\setminus\{l\}$, i.e., each visiting circle is visited by only one viable path;
3. $\max_{l \in [1,k]}\{|P_l|\}$ is minimized, where the function $len(P_l)$ gives the length of P_l.

Here the objective is to minimize the length of the longest path, denoted as k-length. This objective can make the paths have similar lengths. Since the k robots move at the same speed, the collection times of these paths are more or less equal. It easy to understand that this problem is an NP-hard problem. If the nonholonomic constraint is removed, C_i is reduced to s_i. Then it is reduced to the k-TSPN. If $k = 1$, it is further reduced to the TSPN, which is known as NP-hard. So this problem is also NP-hard.

Three relevant approaches have been proposed for the case of multiple robots. The first is called the VRP [2]. Since the VRP framework targets on minimization of the total length of all the paths, it may lead to the situation where the lengths of the produced paths are quite different from each other. The second one is the cluster-based approach which first divides the network into several clusters and then finds the optimal path in each cluster. The cluster-based approach (e.g., K-means [8,23]) is known for effectively decreasing the problem scale, but the path lengths are still not guaranteed to be similar. The third one is the split-based approach (e.g., k-SPLITOUR [15]), which is first to construct a whole path for the single robot and then divide it into several parts by carefully selecting $k - 1$ split positions. It is noted for generating paths of more or less equal lengths. However, we point out that the paths generated by k-SPLITOUR may not be guaranteed to be optimal since it simply connects the initial position to the selected split positions.

Based on these considerations, we propose an algorithm, called k-SVPP. k-SVPP makes use of the profits of the cluster-based and split-based approaches to compensate for their defects. Particularly, we use k-SPLITOUR to guide clustering and find the optimal path in each cluster. The k-SVPP algorithm is given by Algorithm 3; its main steps include: (1) run SVPP to get a whole path; (2) use k-SPLITOUR to split the whole path into k parts and get k clusters; and (3) run SVPP to reconstruct the k paths.

Algorithm 3 k-Shortest viable path planning.

1: Run SVPP to get path P with length $L = |P|$ and permutation $\Sigma = \{\sigma_1, ..., \sigma_n\}$.

2: For each l, $l \in [1, k - 1]$, find the last visiting circle $\sigma_{i(l)}$ along P such that $len(\sigma_1, \sigma_{i(l)}) \leq \frac{l}{k}(L - 2L_{max}) + L_{max}$. Construct k clusters as follows:

$Cluster_1 = \{\sigma_1, \sigma_2, ..., \sigma_{i(1)}\}$,
$Cluster_l = \{\sigma_1, \sigma_{i(l-1)+1}, ..., \sigma_{i(l)}\}, l \in [2, k - 1]$,
$Cluster_k = \{\sigma_1, \sigma_{i(k-1)+1}, ..., \sigma_n\}$.

3: Run SVPP to compute P_l on $Cluster_l$, $l \in [1, k]$.

In Algorithm 3, $L_{max} = \max_{i \in [1,n]} |\mathcal{L}(X_{\sigma_1}, X_{\sigma_i})|$, i.e., the largest length between the configurations on σ_1 and the other visiting circles. Function $len()$ calculates the path length between two given configurations, and $\cup_{l=1}^k Cluster_l = \{C_1, ..., C_n\}$ and $Cluster_{l_1} \cap Cluster_{l_2} = C_1$, $l_1, l_2 \in [1, k]$, $l_1 \neq l_2$. Note the structure of k-SVPP is the same as that of k-SPLITOUR. The main differences between them are as follows. First, k-SPLITOUR is originally designed for the k-TSP while in k-SVPP all the path segments have taken into account the concept of viable path. We can replace all the paths generated by k-SPLITOUR with viable paths and we call the corresponding method viable k-SPLITOUR. Second, in k-SPLITOUR the second (final) step is to construct k permutations, i.e.,

$$\Sigma_1 = \{\sigma_1, \sigma_2, ..., \sigma_{i(1)}\},$$
$$\Sigma_l = \{\sigma_1, \sigma_{i(l-1)+1}, ..., \sigma_{i(l)}\}, l \in [2, k - 1],$$
$$\Sigma_k = \{\sigma_1, \sigma_{i(k-1)+1}, ..., \sigma_n\},$$

while in k-SVPP we conduct a further operation (Step 3), i.e., reconstructing the path in each cluster. Suppose the lengths of the paths generated by viable k-SPLITOUR and k-SVPP are represented by $Length_l$ and $Length_l^*$, $l \in [1, k]$, respectively. We claim that $Length_l \geq Length_l^*$, $\forall l \in [1, k]$. Then we have $\max_l\{Length_l\} \geq \max_l\{Length_l^*\}$, i.e., the k-length of paths by k-SVPP is no greater than that of viable k-SPLITOUR. Behind this, the cost is to run SVPP for another k times to get the k reconstructed paths in k-SVPP, while it is straightforward to construct paths based on the k permutations in viable k-SPLITOUR.

4.5 Simulation results

This section is divided into three parts. The first part demonstrates the performance in some instances and investigates the influence of different factors on SVPP. The second part displays the performance of k-SVPP, and the third part provides a comparison with multihop communication.

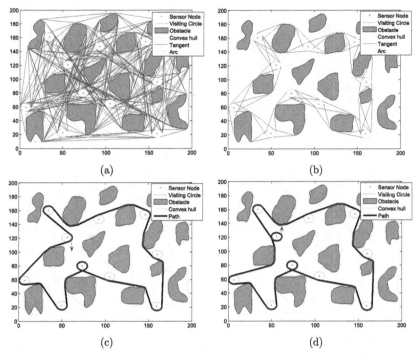

FIGURE 4.4 A demonstrative example with $n = 10$. (a) Tangent graph. (b) Simplified tangent graph. (c) Given a south facing initial heading, the shortest viable path length is 749.1 m. (d) Given a north facing initial heading, the shortest viable path length is 786.7 m.

4.5.1   Performance of SVPP

We simulate a 200 m × 200 m virtual field with a set of disjoint obstacles. In this field, n (10–50) sensor nodes are randomly deployed outside the obstacles. According to the assumption that two elements do not intersect, (1) any two obstacles are at least $2d_{safe}$ from each other; (2) any two sensor nodes are at least $2R_{min}$ from each other; and (3) any sensor node and obstacle are at least $R_{min} + d_{safe}$ from each other. We further assume that the sensor network is connected, i.e., any node has at least one neighbor node within d_{max} distance. This enables the nodes to transmit the data to the base via multihop communication. The sensor nodes are event-driven and the data load g is between 0 and 1 MB. We consider k identical ground robots, whose maximum angular speed u_M is set as 1 rad/s. The speed v is between 1 and 6 m/s. Thus the corresponding minimum turning radius R_{min} is between 1 and 6 m.

This section displays the performance of SVPP. Since the base station is treated in the same way as sensor nodes, we present them using the same mark in the following figures. A network instance with 10 nodes is shown in Fig. 4.4 together with its tangent graph and simplified tangent graph. The speed of the

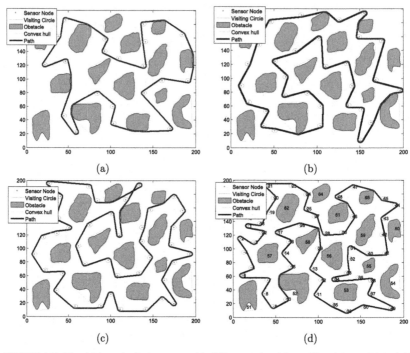

FIGURE 4.5 The viable paths for instances with different scales. (a) $n = 20$ and the path length is 907.0 m. (b) $n = 30$ and the path length is 1078.4 m. (c) $n = 40$ and the path length is 1275.9 m. (d) $n = 50$ and the path length is 1352.4 m.

robot is taken as 6 m/s. Applying the proposed algorithm SVPP, we get the shortest viable path. Here we demonstrate the two shortest viable paths given two different initial headings. We also conduct simulations for the networks with 20, 30, 40, and 50 sensor nodes shown in Fig. 4.5, where $v = 3$ m/s. In these simulations, we set a small value to g such that the data can be loaded shortly.

Next, we investigate the influences of two key parameters, i.e., robot speed v and data load g, on two system metrics: path length and collection time (path length/v). Here $v = 1, 2, 3, 4, 5,$ and 6 m/s and $n = 10, 20, 30, 40,$ and 50. For each pair of v and n, we simulate 20 independent instances; the results are shown in Fig. 4.6. For a fixed n, the path length tends to increase with the increase of v, due to the increase of R_{min}. Extremely, if $v = 0$, it turns to the case of the TSP. We do not display the TSP paths as they are not viable. In Fig. 4.6b, the collection time decreases with increasing v. From these simulations we can see that although increasing the robot speed increases the path length, the collection time can be reduced. Note the data load g still takes a small value. When a large data load is taken into account, such a conclusion may not be appropriate.

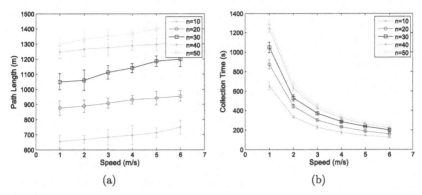

FIGURE 4.6 The impacts of speed on path length and collection time. (a) Path length. (b) Collection time.

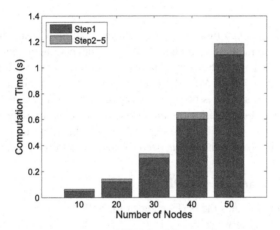

FIGURE 4.7 Average computation time of SVPP for different-scale networks.

On these instances, we measure the computation time of SVPP on a 64-bit Windows machine with an Intel(R) Core(TM) i5-4570CUP @3.20GHz processor. We display the average time for each network scale in Fig. 4.7. It shows that the major time consumption is made on Step 1 to calculate the permutation, while Steps 2–5 to search the path take a relatively short time.

Now we investigate the impact of g on system metrics and we focus on a network such as the one shown in Fig. 4.4. Here, the data loads are uniform among all the nodes, which are between 0 and 1 MB. For each pair of g and v, we compute the shortest viable path by SVPP; the results are shown in Fig. 4.8. We can see that for a fixed v, the path length is nondecreasing with g and it has an initial stabilization. For example, for $v = 3$ m/s, with increasing g, the path length remains stable at first and then starts to increase after $g = 0.08$ MB. The reason behind this is among the nodes' regular contact times. The shortest one allows to transmit up to 0.08 MB. When g is larger than 0.08 MB, as the

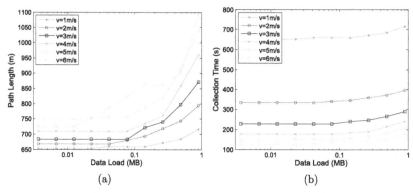

FIGURE 4.8 The impacts of data generation rate on path length and collection time. (a) Path length. (b) Collection time.

shortest regular contact time is not enough, the reading adjustment is applied, which leads to an increased path length. Another interesting phenomenon is that the length of the path with higher v may be shorter than that with lower v. For example, when $g = 0.02$ MB, the path length with $v = 5$ m/s is longer than that with $v = 6$ m/s. This is because the regular contact times on the path with $v = 6$ m/s are still enough for the data load, while several circles are added to the path with $v = 5$ m/s. Fig. 4.8b shows that the collection time increases with data load and increasing the robot speed can reduce the collection time.

4.5.2 Performance of k-SVPP

This section investigates the performance of k-SVPP.

We focus on a network shown in Fig. 4.5d. The node in the central of the field, i.e., s_{29}, is regarded as the base station, and the robots start and end at its visiting circle. We present the results of $k = 2$ and $k = 3$. For comparison, the paths generated by viable k-SPLITOUR are also displayed. For $k = 2$, the split visiting circle is C_{28}. Then $\{C_{29}, C_{26}, ..., C_{28}\}$ forms the first cluster and $\{C_{29}, C_{30}, ..., C_{13}\}$ is the second. In this instance, k-SVPP and viable k-SPLITOUR generate the same paths (Fig. 4.9). For $k = 3$, the two split visiting circles are C_{18} and C_{43}. Then $\{C_{29}, C_{26}, ..., C_{18}\}$ forms the first cluster, $\{C_{29}, C_{19}, ..., C_{43}\}$ forms the second, and $\{C_{29}, C_{42}, ..., C_{13}\}$ is the third. The three paths generated by k-SVPP are shown in Fig. 4.10a and those of viable k-SPLITOUR are demonstrated by Fig. 4.10b. Comparing these two results, we can see that the paths generated by k-SVPP outperform those generated by viable k-SPLITOUR. The k-length of k-SVPP paths is 528.8 m and that of viable k-SPLITOUR is 569.2 m. Thus, in this instance, k-SVPP reduces path length by 7.6%. The reason behind this improvement is obvious, i.e., as discussed in Section 4.4, the split permutations are not optimal in the clusters. Thus, the paths constructed based on these permutations are not the shortest.

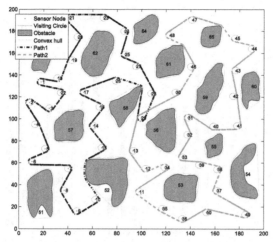

FIGURE 4.9 When $k = 2$, the paths generated by k-SVPP and viable k-SPLITOUR for the instance shown in Fig. 4.5d are the same. The two path lengths are 710.2 m and 688.3 m, respectively.

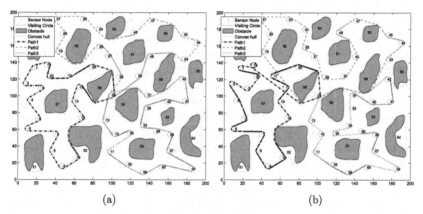

(a)   (b)

FIGURE 4.10 Comparison of k-SVPP and viable k-SPLITOUR when $k = 3$. (a) k-SVPP (path lengths are 528.8 m, 517.2 m, and 517.0 m). (b) k-SPLITOUR (path lengths are 569.2 m, 559.4 m, and 552.8 m).

Further, we apply k-SVPP to more network instances and compare the performance with that of viable k-SPLITOUR for $k = 3$. The results are shown in Fig. 4.11. We can see that k-SVPP performs no worse than viable k-SPLITOUR, and the former yields a 5.5% improvement on average.

4.5.3 Comparison with multihop communication

This subsection compares our work with a multihop communication algorithm: shortest path routing. We focus on energy consumption since it is an important

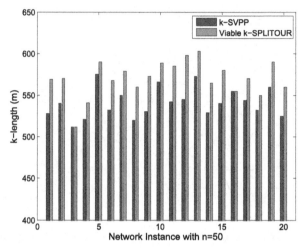

FIGURE 4.11 Comparison of k-SVPP and viable k-SPLITOUR when $k = 3$ for the 20 network instances with $n = 50$ generated in Section 4.5.1.

system metric when the sensor nodes have limited power supplies. We consider a data transmission rate model of two stairs; the corresponding parameters are $r_1 = 250$ KB/s, $r_2 = 19.2$ KB/s, $d_1 = 20$ m, $d_{max} = d_2 = 50$ m [20]. Associated with the transmission rates, the transmission energy consumption rates are $e_1 = 6.8 \times 10^{-6}$ J/bit and $e_2 = 1.1 \times 10^{-5}$ J/bit, respectively.

In shortest path routing, every node transmits its data to the base through the path with the shortest distance (such shortest path can be found by existing algorithms such as Dijkstra's algorithm [24]). Since transmitting data usually consumes more energy than sensing and receiving, the energy consumption of the network using multihop communication can be simply expressed as

$$E_{multihop} = \sum_{i}^{n} eh_i g_i, \tag{4.15}$$

where h_i is the hop number from s_i to the base, g_i is the data load of s_i, and e indicates the energy consumption for transmitting one-bit data. More details about (4.15) can be found in [25]. If we assume the distance between any pair of nodes is larger than d_1, then $e = e_2$. Considering R_{min}, the sensor nodes can transmit data to the robots via a single hop and the energy consumption rate is e_1. We use $E_{singlehop}$ to represent the energy consumption when robots are used:

$$E_{singlehop} = E_{node} + E_{robot}, \tag{4.16}$$

where $E_{node} = \sum_{i}^{n} e_1 g_i$ and $E_{robot} = \lambda L$ [26]. Now L gives the total path length of all the robots and λ depends on the robot and its speed. We will see the value of λ influences the performance. Here, λ is set to be 0.035.

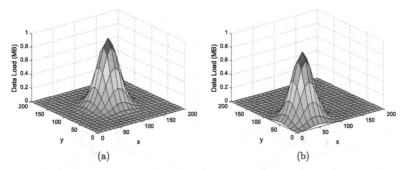

FIGURE 4.12 Distributed data loads from a multivariate Gaussian model where the covariance matrix is [400, 0; 0, 400]. The node nearest to the position of the event generates up to 1 MB data during $\mathcal{T}$. (a) Central. (b) Southwest.

In the following part, we focus on the network instances of 50 nodes shown in Sections 4.5.1 and 4.5.2. Unlike Section 4.5.1, where we assume uniform data loads, the data loads in this part are distributed in the network, which is more practical. By a multivariate Gaussian model, we generate the distributed data loads as shown in Fig. 4.12. We study the following four data load distributions. (1) Central: an interested event occurs at position (100, 100) of the field (Fig. 4.12a). The closer a node to this position, the more data it generates. (2) Southwest: an interested event occurs at position (50, 50) (Fig. 4.12b). Also, we consider two random distributions. (3) R-low: the data loads across the network are generated *randomly* at a *low* level (0–0.5 MB). (4) R-high: the data loads are generated *randomly* at a *high* level (0.5–1 MB).

We apply SVPP, k-SVPP, SVPP-no-adjustment, shortest path routing, and viable k-SPLITOUR on the network shown in Fig. 4.5d; the results are displayed in Fig. 4.13. Note that the results are mean values of the 20 independent instances.

Fig. 4.13a demonstrates the energy consumption by sensor nodes and robots to collect data in these four situations. We learn the following things. The energy consumption by the shortest path routing algorithm for southwest is higher than for central distributions. The reason is that as the base station is located in the central area, the sensor nodes in the southwest case consume more energy on relaying data, although the total data loads of all the nodes are similar in these two cases. The shortest path routing algorithm spends the most energy for R-high and the least for R-low distributions due to the total amount of data loads. For the algorithms using robots, SVPP-no-adjustment and SVPP employ a single robot while k-SVPP and viable k-SPLITOUR both employ three robots. Fig. 4.13a shows that using a single robot consumes less energy than using multiple robots. This is reasonable since the total path length of the multiple robots is longer than in the single robot case, as shown in Section 4.5.2. The benefit of employing multiple robots is that the collection time is shortened. Among the two schemes using a single robot, SVPP costs more energy than SVPP-no-

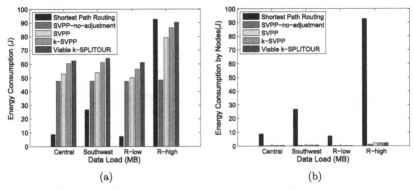

FIGURE 4.13 Energy consumption by shortest path routing, SVPP without adjustment, SVPP, k-SVPP, and viable k-SPLITOUR. (a) By sensor nodes and robots. (b) By sensor nodes.

adjustment. The excess part is used on the rotating movement. We calculate the average collected data percentage of SVPP-no-adjustment in these four distributions: 55%, 62%, 95%, and 53%. To collect all the data, the robot needs to do at least one more tour along the paths. Then SVPP-no-adjustment spends more energy than SVPP. Among the two schemes using multiple robots, viable k-SPLITOUR costs more than k-SVPP. The reason is the same as that discussed in Section 4.5.2. Comparing these schemes on the four distributions, shortest path routing consumes the least energy in the first three cases, since robot movement generally consumes more energy. To exclude the influence of movement energy consumption, we display the part consumed by sensor nodes only in Fig. 4.13b. It shows that in every case, using robots to collect data saves sensor nodes around 95% energy compared to multihop communication. Also, in terms of total energy consumption, using multihop communication may spend more energy when the data loads are large, such as the R-large case in Fig. 4.13a.

We indicate that the results shown in Fig. 4.13a depend on λ. Increasing this factor may lead to large energy consumption by the schemes using robots. Although using robots may consume more than conventional multihop communication in terms of total energy consumption, the approaches using robots are still promising since the energy consumption by nodes can be saved prominently, as shown in Fig. 4.13b; besides, the robots can be recharged much more easily than the sensor nodes.

4.6 Discussion

In this section we provide a short discussion of our work.

4.6.1 Extension

Our proposed algorithms SVPP and k-SVPP are designed for the data collection systems with two layers: a sensor layer and a base station layer. But both of them are not restricted to systems of two layers. They can be extended to systems of three layers, i.e., a sensor layer, a gateway layer, and a base station layer, by adding inner network communication protocols between sensor and gateway layers. The sensor nodes can be clustered locally into different groups and use inner network communication to transmit the extracted data to their corresponding gateways. Then our algorithms will serve the gateway and base station layers in the same way as we do above. As recent studies on the area of WSNs have proposed many network communication protocols, extending our work to three-layer systems will be efficient and effective.

4.6.2 Limitation

We do not consider the computation capability of robots. In other words, we assumed that obstacles will be avoided when following the designed paths. However, this is based on the accuracy of the given information of obstacles such as shapes and positions. We note that in reality such information may be inaccurate or unavailable, for example when movable obstacles exist in the sensing field. In this scenario, the proposed algorithms would not be appropriate.

4.6.3 Application

The proposed work can be applied to applications using wheeled robots with bounded turning radius to collect data from sensor nodes. It may save sensor nodes much energy in the scenario of large data loads, for example in wireless multimedia sensor networks.

The proposed algorithms are originally designed for the applications of collecting information from sensor nodes. They can also be applied in a reverse manner, i.e., to distribute information to the sensor nodes, such as charging the sensor nodes [27], due to the availability of the wireless charging technique. The charging amount can be regarded as the data load, which is determined by the residual energy in the node and its capacity.

4.7 Summary

This chapter studies the problem of planning the shortest viable path for unicycle robots with bounded turning radius serving as data collectors in a cluttered sensing field. We define a viable path that combines the concerns of both robotics and sensor networks. In many applications, this term is closer to reality. We formulate the problem of planning the shortest viable path for a single robot as a variant of the DTSPN. Accordingly, we develop an SVPP algorithm. We further consider the problem of planning viable paths for multiple robots and present a k-SVPP algorithm. We conduct simulations with different network scales, robot speeds, and distributed data loads to show the performance of the proposed algorithms. Also, comparisons with existing alternatives are provided. We find that SVPP and k-SVPP are effective to design viable paths for unicycle robots with bounded angular speed. Compared to multihop communication, our algorithms can save around 95% energy for sensor nodes. Furthermore, increases in robot speed and the employment of multiple robots can reduce the data collection time significantly. In this chapter, we consider several practical issues existing in the utilization of mobile robots to collect data, and the presented results are meaningful for real-world applications.

Moreover, we apply our approaches to the problem of point to point navigation, with the background of safe mission planning. We propose an optimization model to navigate an aircraft or a flying robot to its final destination while minimizing the maximum threat level and the length of the aircraft path. The construction of optimal paths involves a simple geometric procedure and is very computationally efficient. The effectiveness of the proposed method has been demonstrated by illustrative examples and comparisons with existing work. It should be pointed out that we consider a 2D or planar navigation problem. An important direction of future research will be an extension of the presented planar algorithm to practically important cases of 3D threat environments.

References

[1] E.L. Lawler, The Traveling Salesman Problem: a Guided Tour of Combinatorial Optimization, Wiley Interscience Series in Discrete Mathematics, 1985.
[2] G.B. Dantzig, J.H. Ramser, The truck dispatching problem, Management Science 6 (1) (1959) 80–91.
[3] L.E. Dubins, On curves of minimal length with a constraint on average curvature, and with prescribed initial and terminal positions and tangents, American Journal of Mathematics (1957) 497–516.
[4] A.S. Matveev, H. Teimoori, A.V. Savkin, A method for guidance and control of an autonomous vehicle in problems of border patrolling and obstacle avoidance, Automatica 47 (3) (2011) 515–524.
[5] A.S. Matveev, H. Teimoori, A.V. Savkin, Navigation of a unicycle-like mobile robot for environmental extremum seeking, Automatica 47 (1) (2011) 85–91.
[6] A.S. Matveev, H. Teimoori, A.V. Savkin, Range-only measurements based target following for wheeled mobile robots, Automatica 47 (1) (2011) 177–184.
[7] A.V. Savkin, M. Hoy, Reactive and the shortest path navigation of a wheeled mobile robot in cluttered environments, Robotica 31 (2) (2013) 323–330.
[8] A. Wichmann, T. Korkmaz, Smooth path construction and adjustment for multiple mobile sinks in wireless sensor networks, Computer Communications 72 (2015) 93–106.
[9] A.K. Kumar, K.M. Sivalingam, Energy-efficient mobile data collection in wireless sensor networks with delay reduction using wireless communication, in: The 2nd International Conference on Communication Systems and Networks (COMSNETS), IEEE, 2010, pp. 1–10.
[10] G. Xing, T. Wang, W. Jia, M. Li, Rendezvous design algorithms for wireless sensor networks with a mobile base station, in: The 9th ACM International Symposium on Mobile Ad Hoc Networking and Computing, ACM, 2008, pp. 231–240.
[11] A. Somasundara, A. Ramamoorthy, M.B. Srivastava, et al., Mobile element scheduling for efficient data collection in wireless sensor networks with dynamic deadlines, in: Real-Time Systems Symposium, IEEE, 2004, pp. 296–305.
[12] L. Song, D. Hatzinakos, Architecture of wireless sensor networks with mobile sinks: sparsely deployed sensors, IEEE Transactions on Vehicular Technology 56 (4) (2007) 1826–1836.
[13] K.J. Obermeyer, P. Oberlin, S. Darbha, Sampling-based roadmap methods for a visual reconnaissance UAV, in: AIAA Conference on Guidance, Navigation and Control, 2010.

[14] J.T. Isaacs, J.P. Hespanha, Dubins traveling salesman problem with neighborhoods: a graph-based approach, Algorithms 6 (1) (2013) 84–99.

[15] G.N. Frederickson, M.S. Hecht, C.E. Kim, Approximation algorithms for some routing problems, SIAM Journal on Computing 7 (1978) 178–193.

[16] H. Huang, A.V. Savkin, Viable path planning for data collection robots in a sensing field with obstacles, Computer Communications 111 (2017) 84–96.

[17] I.R. Manchester, A.V. Savkin, F.A. Faruqi, Method for optical-flow-based precision missile guidance, IEEE Transactions on Aerospace and Electronic Systems 44 (3) (2008) 835–851.

[18] A.V. Savkin, H. Teimoori, Bearings-only guidance of a unicycle-like vehicle following a moving target with a smaller minimum turning radius, IEEE Transactions on Automatic Control 55 (10) (2010) 2390–2395.

[19] D.J. Struik, Lectures on Classical Differential Geometry, Courier Corporation, 2012.

[20] X. Ren, W. Liang, W. Xu, Data collection maximization in renewable sensor networks via time-slot scheduling, IEEE Transactions on Computers 64 (7) (2015) 1870–1883.

[21] P. Vana, J. Faigl, On the Dubins traveling salesman problem with neighborhoods, in: IEEE/RSJ International Conference on Intelligent Robots and Systems (IROS), IEEE, 2015, pp. 4029–4034.

[22] A.M. Frieze, G. Galbiati, F. Maffioli, On the worst-case performance of some algorithms for the asymmetric traveling salesman problem, Networks 12 (1) (1982) 23–39.

[23] Z. Tang, U. Ozguner, Motion planning for multitarget surveillance with mobile sensor agents, IEEE Transactions on Robotics 21 (5) (2005) 898–908.

[24] E.W. Dijkstra, A note on two problems in connexion with graphs, Numerische Mathematik 1 (1) (1959) 269–271.

[25] S. Gao, H. Zhang, S.K. Das, Efficient data collection in wireless sensor networks with path-constrained mobile sinks, IEEE Transactions on Mobile Computing 10 (4) (2011) 592–608.

[26] F. El-Moukaddem, E. Torng, G. Xing, S. Kulkarni, Mobile relay configuration in data-intensive wireless sensor networks, IEEE Transactions on Mobile Computing 12 (2) (2013) 261–273.

[27] H. Dai, X. Wu, G. Chen, L. Xu, S. Lin, Minimizing the number of mobile chargers for large-scale wireless rechargeable sensor networks, Computer Communications 46 (2014) 54–65.

Chapter 5

Data collection in wireless sensor networks by ground robots with fixed trajectories[☆]

5.1 Motivation

As mentioned in Chapter 4, one side effect of multihop communication to transmit data packets to the static base sinks (BSs) is the funneling effect [1]. Recent studies have shown that using mobile sinks (MSs) to collect data in WSNs can relieve the funneling effect issue [2,3]. An MS traversing the sensing field can collect data from sensor nodes over a short-range communication link [4,5], and then the onboard MS transmits the collected data wirelessly to a remote center, since it has no energy limitation. Long-hop relaying is not used at sensor nodes and the energy consumption is reduced. Traversing the sensing field by MS needs to be timely and efficient because failure to visit some parts of the field leads to data loss, and infrequently visiting some areas results in a long delivery delay. Besides, the trajectory planning of MSs in these cases becomes more difficult. Furthermore, in urban areas, the planned trajectory sometimes cannot be realized since the MS is constrained to roads. Alternatively, amounting an MS on a vehicle, such as a bus, avoids some difficulties and can provide better performance for data collection. First, since the bus is already a component of the environment and its trajectory is predefined, the difficult path planning and complex control of the MS's movement are avoided. Second, instead of visiting each sensor node individually, which is a time-consuming task due to the low physical speed of MSs, combining multihop communication with path-constrained MS can increase the data delivery delay.

This chapter investigates using an MS, which is attached to a bus, to collect data in WSNs with nonuniform node distribution. Such WSNs exist in many applications. For example, in the case of monitoring the air pollution of a city, industrial areas are usually deployed with more sensors than the residential areas. Also, since the areas of interest may be isolated from each other, conventional data collection approaches are not appropriate due to the limited budget

---

☆ The main results of the chapter were originally published in Hailong Huang, Andrey V. Savkin, An energy efficient approach for data collection in wireless sensor networks using public transportation vehicles, AEÜ. International Journal of Electronics and Communications 75 (2017) 108–118. Permission from Elsevier for reuse was obtained.

Copyright © 2022 Elsevier Inc. All rights reserved.

of energy resource. In this case, an MS mounted on a bus is able to relieve the bottleneck of energy at sensor nodes, because the MS can serve the isolated areas at different times, as if there is a virtual static sink for each area and such sink only works at a specified time duration, i.e., the duration during which the MS is in the area. Instead of the coverage problem studied in publications [6–8], the focus here is on routing the sensory data from source nodes to the MS in an energy-efficient way such that the energy expenditure is balanced across the entire network.

The main contributions of this chapter are a clustering algorithm and a routing algorithm. The core of the clustering algorithm lies in the selection of CHs. Intending to balance energy consumption, we design unequal cover ranges for CHs considering the feature of nonuniform node distribution. The cover range of a CH depends on its distance to the MS and the local node density. Unlike other works, the distance here is the hop distance instead of the Euclidean distance. Removing the ability of measuring Euclidean distance simplifies the sensor nodes. The unequal cover ranges can make the clusters with similar distances to the MS have approximate sizes such that the energy consumption by CHs can be balanced. Since the designed cover range does not exceed the single-hop communication range, the cluster members (CMs) consume energy approximately also. The proposed routing algorithm associates each CH to a CH that is closer to the MS's trajectory. The CH u will associate to a CH v only if v's hop distance is no larger than u's. Furthermore, the association accounts for the residual energy of CH v and the number of attached CMs. We compare our approach with some existing ones through simulations and we conclude that our approach achieves a longer network lifetime.[1] The main results of the chapter were originally published in [9].

The remainder of this chapter is organized as follows. Section 5.2 presents the network model. Section 5.3 discusses the proposed protocol in details, which is followed by some theoretical analysis in Section 5.4. Section 5.5 provides extensive simulations to evaluate the proposed approach. Finally, Section 5.6 summarizes this chapter.

5.2 Network model

Consider a WSN consisting of n static sensor nodes nonuniformly deployed in the field. A bus carrying an MS moves the predefined trajectory following its timetable. We consider the following assumptions.

1. The nodes, as well as the MS, have unique IDs.
2. All the nodes use power control to adjust the transmitting power.
3. If a node works as CH, it aggregates the received data packets within a cluster into one packet, while it does not aggregate the data packets from other CHs.

---

[1] There are several definitions of network lifetime in the literature. Here we adopt the definition of network lifetime as the number of rounds until the first node exhausts its energy reserve, which has been widely used.

Furthermore, the raw data packets and the aggregated packets have the same size.

We consider the energy dissipation model used in previous work, e.g., [10–15]:

$$E_t(l, d) = \begin{cases} l \times E_{elec} + l \times E_{fs} \times d^2, & if \ d \le d_0, \\ l \times E_{elec} + l \times E_{mp} \times d^4, & if \ d > d_0, \end{cases} \tag{5.1}$$

where $E_t(l, d)$ is the total energy dissipated to deliver a single l-bit packet from a transmitter to its receiver over a single link of distance d. The electronic energy E_{elec} depends on electronic factors such as digital coding, modulation, filtering, and spreading of the signal. The amplifier energy in free space E_{fs} or in a multipath environment E_{mp} depends on the distance from the transmitter to the receiver, and the threshold is d_0.

For receiving data packets, the sensor nodes expand energy according to

$$E_r(l) = l \times E_{elec}. \tag{5.2}$$

Note that data packet transmission and control message exchange both follow models (5.1) and (5.2).

Additionally, we assume that the energy consumption for sensing and data aggregation are, respectively,

$$E_s(l) = l \times E_{sens} \tag{5.3}$$

and

$$E_a(l) = l \times E_{aggr}, \tag{5.4}$$

where E_{sens} depends on electronic factors and E_{aggr} relates to the aggregation algorithm.

5.3 Routing protocol

The proposed protocol scheme contains two stages: initial and collection stages. Fig. 5.1 illustrates the protocol operation by the timeline. Basically, the initial stage aims at making every node aware of the information required to operate the following procedures. The collection stage consists of a number of data collection cycles. At the beginning of each cycle, the network constructs cluster formation. In this phase, the sensor nodes transmit a control message to their neighbor nodes and build up a network structure in a distributed manner. Then a certain number of data collection rounds are operated. The control messages used here are described in Table 5.1.

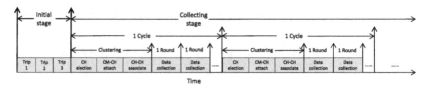

FIGURE 5.1   The operation of the proposed protocol by timeline.

TABLE 5.1 Description of control messages.

Message	Description
Initial-Msg	hop, ID
Return-Msg	hop
Hop-Msg	h_{max}, h_{min}
CH-Compete-Msg	Energy, ID
CH-Win-Msg	ID
CH-Quit-Msg	ID
CM-CH-Offer-Msg	ID
CM-CH-Request-Msg	ID
CM-CH-Confirm-Msg	ID
CH-CH-Offer-Msg	Energy, CM size, hop, ID
CH-CH-Confirm-Msg	ID
Data-Request-Msg	ID

5.3.1   Initial stage

The initial stage requires the MS to make three trips on its path. The purpose here is to get the hop distance to the MS's trajectory as well as the local node density for each node. Such information plays a significant role in the collection stage.

Trip 1. When the MS moves, it continuously broadcasts Initial-Msg containing MS ID and hop distance (the hop distance equals 0). A node, which is within the communication range of the MS, receives the packet and executes the following procedures to extract some information from the message and modify it: (1) extracting the ID in the packet as its parent node ID; (2) replacing the ID with its own ID; (3) increasing the hop distance by 1; and (4) extracting the hop distance as its hop distance to MS. Then it broadcasts the modified message to the nodes within its communication range. Note that it is possible for one node to receive more than one message. If the hop distances are different, it selects the smallest one (see Fig. 5.2 for an example); otherwise, it selects the most early received one. At the same time, it keeps the number of received messages as its neighbor count. At the end of this trip, every node i knows its hop distance (h_i) to the MS, the parent node, as well as the neighbor count (n_i).

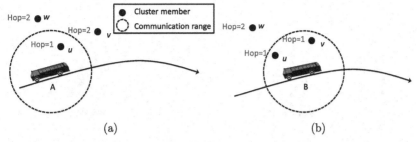

FIGURE 5.2 Hop distance construction. When the MS is at A, u is within range. The hop distance of u is 1. Since v and w are within the range of u, their hop distances are 2. When the MS is at B, v is within the range of the MS, and then the hop distance is changed to 1. Note that w never comes into the range of the MS, but it is within the range of u, thus its hop distance remains 2.

Trip 2. Every sensor node transmits Return-Msg containing its hop distance to the MS to its parent node. On this trip, the MS keeps receiving packets from the nodes nearby. At the end of Trip 2, the MS knows the maximum and minimum hop distance (h_{max} and h_{min}) from sensor nodes to itself.

Trip 3. The MS keeps broadcasting Hop-Msg containing h_{max} and h_{min}. The nodes receive such message extract h_{max} and h_{min} and forward the packet to other nodes within range. Finally, every node in the network knows h_{max} and h_{min}.

5.3.2 Collection stage

The collection stage is the main stage of our protocol. It consists of a number of data collection cycles. At the beginning of each cycle, the sensor nodes reorganize themselves by constructing new clusters. After that, CMs send the sensory data to CHs and CHs send the aggregated data to the MS directly or to another CH for relay. When the bus finishes its trip, we say one round of data collection is completed. Thus, each cycle consists of clustering and k rounds of data collection. Obviously, the clustering result plays a significant role in the following data collection since it impacts the energy consumption of both CHs and CMs. The parameter k also influences the energy consumption. A small value for k means that the network needs to reconstruct the clusters frequently, resulting in a large number of control messages to exchange. On the other hand, a large value for k makes the network operate under one cluster structure for a long time, which may lead to the phenomenon that some CHs cannot survive the current cycle. Below, we describe the main phases in the collection stage in detail.

5.3.2.1 CH election

For the CH election, we extend the method in [10]. In [10], the authors consider the scenario that the sensor nodes are uniformly deployed and design each CH's cover range based on its Euclidean distance to the static BS. In contrast, we con-

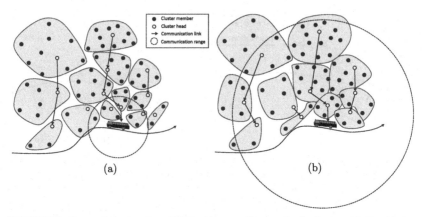

FIGURE 5.3 Cluster formation for a WSN with nonuniform node distribution. (a) Considering node distribution. (b) Not considering node distribution. When node distribution is not considered, CHs having a similar distance to the MS's trajectory have a similar cover range. Thus, the CHs in dense areas have more CMs than those in sparse areas.

sider the nonuniform deployment of sensor nodes, and we do not use Euclidean distance information since such information may not be reliable in a harsh environment and measuring such information is costly. Instead, we use hop distance to calculate the cover range. The key equation to compute the cover range of each CH is

$$R_i = (1 - \alpha \frac{1}{\sqrt{\rho_i}} \frac{h_{max} - h_i}{h_{max} - h_{min}})R_0, \qquad (5.5)$$

where R_i, ρ_i, and h_i are respectively the cover range, local node density, and hop distance to the MS of CH i, α is a given positive constant, and R_0 is the communication range of the sensor nodes. Note that ρ_i is the relative node density, which is estimated based on the number of neighbor nodes, i.e., $\rho_i = n_i/\pi R_0^2/\bar{\rho}$. Here $\bar{\rho}$ is the average node density of the sensor network and can be obtained in the node placement phase.

The fundamental idea to design the cover range like (5.5) is as follows. To make two CHs with the same hop distance to the MS have an approximately equal number of CMs, we try to make $\pi R_i^2 \rho_i = \pi R_j^2 \rho_j$ hold, from which we obtain $R_i \propto 1/\sqrt{\rho_i}$, i.e., the CH in the dense area has a smaller cover range while the CH in the sparse area has a larger cover range (see Fig. 5.3 for an example). Besides, consistent with [10], the larger the hop distance to the MS is, the larger the cover range will be. Thus, $R_i \propto h_i$. Eq. (5.5) works for both scenarios of uniform and nonuniform node distributions. For a uniform node distribution, node densities are approximately equal across the entire network, thus $1/\sqrt{\rho_i}$ influences R_i little.

With this cover range in place, the network is in the phase of CH election. A set of CH candidates is first elected. A node i randomly generates $\mu \in (0, 1)$ [16]. Let e_i be node i's residual energy. If $\mu e_i > \beta$ (β is the given threshold), node i becomes a CH candidate; otherwise, it becomes a CM. The CH candidates turn into a CH election phase. In the CH election phase, every CH candidate broadcasts CH-Compete-Msg containing its residual energy and ID within its cover range. The other CH candidates within such cover range receive the message and compare the contained residual energy with their own. The competition rules are as follows. The CH candidates with the highest residual energy within their cover range become CHs. The CHs broadcast CH-Win-Msg to inform their status. The CH candidates that receive CH-Win-Msg quit the CH competition process and broadcast CH-Quit-Msg, because receiving CH-Win-Msg means they are covered by at least one CH's cover range. A CH candidate that has not received CH-Win-Msg listens for CH-Quit-Msg. Once received CH-Quit-Msg, it ignores the corresponding node and checks whether its residual energy is the largest in the left set. This process lasts until timer $T_{election}$ expires. The CH election algorithm is given as Algorithm 1.

Algorithm 1 CH election.

1:    $\mu \longleftarrow rand(0, 1)$
2:    **if** $\mu e_i > \beta$ **then**
3:       Broadcast CH-Compete-Msg within R_i
4:       Listen for CH-Compete-Msg
5:       Find the message which contains the largest residual energy e_j
6:       **while** timer $T_{election}$ has not expired **do**
7:          **if** $e_j < e_i$ **then**
8:             Broadcast CH-Win-Msg within R_i; exit
9:          **end if**
10:       Listen for CH-Win-Msg
11:       **if** CH-Win-Msg is received **then**
12:          Broadcast CH-Quit-Msg within R_i; exit
13:       **end if**
14:       Listen for CH-Quit-Msg
15:       **if** CH-Quit-Msg is received **then**
16:          Get rid of the message corresponding to the ID in CH-Quit-Msg
17:          From the left messages, find the one which contains the largest residual energy e_j
18:       **end if**
19:       **end while**
20: **end if**

5.3.2.2 CM-CH attachment

After the CH election, every CM needs to attach to a CH. The CM-CH attachment algorithm is shown as Algorithm 2. In this phase, CH and CM send a message or hear alternatively. At the beginning, CH broadcasts a CM-CH-Offer-Msg within its cover range. Once a CM hears a CM-CH-Offer-Msg, it sends a

---

Algorithm 2 CM-CH attachment.

1: **if** node i is a CH **then**
2:     Broadcast CM-CH-Offer-Msg within R_i.
3:     Listen for CM-CH-Request-Msg until timer T_{asso} expires
4:     Send CM-CH-Confirm-Msg to the CMs
5: **else**
6:     Listen for CM-CH-Offer-Msg until timer T_{offer} expires.
7:     **if** CM-CH-Offer-Msg is received **then**
8:         Send CM-CH-Request-Msg to the CH
9:         Listen for CM-CH-Confirm-Msg until timer $T_{confirm}$ expires
10:     **else**
11:         **while** CM-CH-Confirm-Msg not received **do**
12:             Increase transmitting range and broadcast CM-CH-Request-Msg
13:             Listen for CM-CH-Confirm-Msg until timer $T_{confirm}$ expires
14:         **end while**
15:     **end if**
16: **end if**

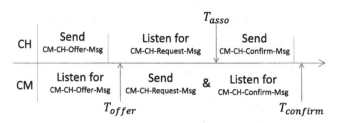

FIGURE 5.4 The operations of CH and CMs in CM-CH attachment.

CM-CH-Request-Msg back to the CH. Then, CH sends CM-CH-Confirm-Msg to the CMs from which it hears the CM-CH-Request-Msg. If a CM cannot hear CM-CH-Offer-Msg from any CHs, it broadcasts CM-CH-Request-Msg until it joins a CH as shown by lines 11–14 and the initial transmitting range is its own cover range. At the end of this phase, any CH (CM) knows its associated CMs (CH). We introduce several timers to support the implementation of CM-CH attachment. CM listens for CM-CH-Offer-Msg until timer T_{offer} expires and CH listens for CM-CH-Request-Msg until timer T_{asso} expires. After timer T_{offer} expires, no matter a CM receives CM-CH-Offer-Msg or not, it starts to send CM-CH-Request-Msg and listen for CM-CH-Confirm-Msg until timer $T_{confirm}$ expires. We display the operations of CH and CM in this phase in Fig. 5.4.

For intracluster data collection, the TDMA schedule is used. The CH sets up a TDMA schedule based on the number of its CMs and transmits it back to its

CMs. After the TDMA schedule is known by all CMs in the cluster, the CM-CH attachment phase is completed.

5.3.2.3 CH-CH association

The next phase of the protocol is CH-CH association. The objective of CH-CH association is to find a route for each CH such that it can transmit its data to the MS. Each CH broadcasts CH-CH-Offer-Msg containing its residual energy, the number of CMs attached to it (henceforth called CM size), its ID, and the number of hops l within a range of γR_0 ($\gamma \geq 1$ is a parameter to adjust the transmission range). The CH with a higher hop number, e.g., $l + 1$, and hearing such message selects the one with the largest η (η = residual energy/CM size) and replies a CH-CH-Confirm-Msg to the sender. The consideration behind this is that since the nodes are nonuniformly deployed, the same CM sizes of two CHs cannot be guaranteed. Thus, the CH having a small CM size is able to perform more data relaying tasks. Note that the CH having a lower hop number, e.g., $l - 1$, may also hear the CH-CH-Offer-Msg from the senders with hop number l, but it will simply ignore the message. The CH-CH association algorithm is shown as Algorithm 3. Note that the CHs with hop number 1 do not accept CH-CH-Offer-Msg as those with hop number l, $2 \leq l \leq h_{max}$, so Steps 3–13 of Algorithm 3 are skipped.

Algorithm 3 CH-CH association, executed by CH i.

1: $\eta \leftarrow 0, \text{ID} \leftarrow 0$
2: Broadcast CH-CH-Offer-Msg within γR_0
3: **if** $h_i > 1$ **then**
4:     Listen for CH-CH-Offer-Msg until T_{offer} expires
5:     **for** each CH-CH-Offer-Msg received from CH j **do**
6:         **if** $\eta < j$'s residual energy/CM size **then**
7:             $\eta \leftarrow j$'s residual energy/CM size
8:             $\text{ID} \leftarrow j$'s ID
9:         **end if**
10:     **end for**
11:     Send CH-CH-Confirm-Msg to the CH with ID
12:     Listen for CH-CH-Confirm-Msg
13: **end if**

After all the preparation procedures, the final task is to send data packets to CHs and deliver the buffered data from the CHs near the MS's trajectory to the MS. Different from delivering data to a static sink, the connection of a CH with an MS varies since the MS is moving. To assist the CHs to start and stop transmitting data packets to the MS, the MS periodically broadcasts Data-Request-Msg. The CH receiving Data-Request-Msg transmits data packets to the MS; otherwise, it will not transmit data packets.

5.4 Protocol analysis

In this section, we present the analysis of the proposed protocol. Since the clusters and routing paths are constructed based on control message exchange, we first discuss the message complexity.

We consider the message complexity in terms of sensor nodes instead of MS. In the initial stage, all the sensor nodes forward Initial-Msg, transmit Return-Msg, and forward Hop-Msg. The messages add up to $n + n + n$, i.e., the message complexity for the initial stage is $O(n)$. In the collection stage, the worst case for the CH election is that all n sensor nodes are CH candidates. In such a case, each node broadcasts a CH-Compete-Msg. Suppose m CHs are elected. Then these CHs broadcast CH-Win-Msg, while the others broadcast CH-Quit-Msg. In the phase of CM-CH attachment, m CM-CH-Offer-Msgs are transmitted first and then the other $n - m$ CMs send CM-CH-Request-Msgs. Third, $n - m$ CM-CH-Confirm-Msgs are returned. In the phase of CH-CH association, m CHs broadcast CH-CH-Offer-Msg. Since the CHs with hop number 1 can directly communicate with the MS, they do not need to accept any CH-CH-Offer-Msg. Then, at most m CH-CH-Confirm-Msgs are transmitted back. Therefore, in the worst case, the messages add up to

$$\underbrace{n + m + (n - m)}_{\text{CH election}} + \underbrace{m + 2(n - m)}_{\text{CM-CH}} + \underbrace{2m}_{\text{CH-CH}} = 4n + m, \qquad (5.6)$$

i.e., the message complexity is $O(n)$.

Now we consider the distribution of CHs. Due to Algorithm 1, every CH candidate broadcasts a CH-Compete-Msg within its cover range. Any other CH candidates within such range can receive this message. The one with the highest residual energy wins the competition and it broadcasts CH-Win-Msg. The CH candidates that receive such a message quit the competition by broadcasting CH-Quit-Msg. As mentioned in Section 5.3.2, in the competition phase, only the CH candidates that have the highest residual energy within their own cover ranges can directly decide to be CHs, while all the other CH candidates whose residual energies are not the largest have to wait for either CH-Win-Msg or CH-Quit-Msg from their neighbor CH candidates (see Fig. 5.5 for an example). When CH candidates u, v, and w first exchange CH-Compete-Msg, only v can decide to be CH, while both u and w need to wait for further messages. Then, v broadcasts CH-Win-Msg and u receives such message and broadcasts CH-Quit-Msg. Finally, w receives CH-Quit-Msg. Since w has only neighbor u, it decides to be CH. At the end of this competition, within the cover ranges of v and w, there are no other CHs. To make the conclusion generally, after the CH competition process, it is impossible that two CHs are within each other's cover range.

Another feature of the proposed protocol is that every sensor node is covered by exactly one CH. This feature can be obtained from the phase of CM-CH attachment. On the one hand, if a sensor node is a CH, it is covered by itself.

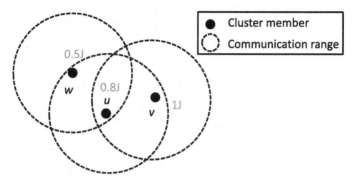

FIGURE 5.5 CH competition process. Here v and w win the competition and become CHs.

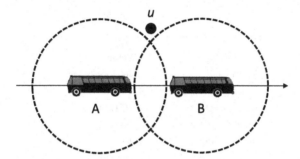

FIGURE 5.6 MS broadcasts Initial-Msg when it is at A and it broadcasts the next Initial-Msg when it arrives at B. Thus, node u fails to hear such message.

On the other hand, if it is a CM, it listens for CM-CH-Offer-Msg and then joins the CH which sends the message. Since the CHs are elected based on a random manner, there are cases where one or several CMs cannot hear CM-CH-Offer-Msg within T_{offer}. In such cases, the CM proactively sends CM-CH-Request-Msg until it hears CM-CH-Confirm-Msg, and then it joins the CH. In this way, the CHs are able to cover all the CMs.

Finally, we discuss the implementation of our protocol. Since the hop distance from a node to the MS's trajectory influences the cover range, our protocol requires all the nodes to have the correct hop distances. The period for broadcasting Initial-Msg impacts on the hop distance construction. If a node that should receive Initial-Msg is not covered by two successful broadcasts, it may get the wrong hop distance. An example is demonstrated in Fig. 5.6. We note that this issue is related to the localization of sensor nodes, the broadcasting frequency, and the MS's speed. Since many applications involve random deployment and the MS is not energy-constrained in our scenario, to address this issue, we assume the MS broadcasts at a high frequency such that all the nodes should hear Initial-Msg correctly, instead of making other assumptions.

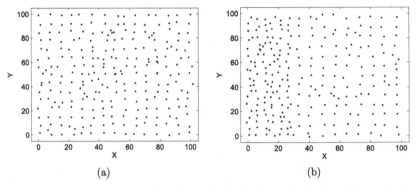

FIGURE 5.7 Topology in each scenario. (a) Uniform node distribution. (b) Nonuniform node distribution.

TABLE 5.2 Parameters of simulations.

Parameter	Value
Sensor field	100 m × 100 m
Number of nodes	250
Data packet size	4000 bits
Control message size	400 bits
Initial energy	1 J
E_{elec}	50 nJ/bit
E_{fs}	10 pJ/(bit m^2)
E_{mp}	0.0013 pJ/(bit m^4)
E_{sens}	1 nJ/bit
E_{aggr}	5 nJ/(bit packet)
d_0	87 m

5.5 Simulation results

We evaluate the performance of our protocol by simulations under two scenarios: Scenario 1, uniform node distribution as shown in Fig. 5.7a, and Scenario 2, nonuniform node distribution as shown in Fig. 5.7b. In Scenario 1, the nodes have similar local densities. In contrast, in Scenario 2, the nodes in the left part have higher densities than the nodes in the right part. We simulate the proposed protocol using MATLAB® with the networks shown in Fig. 5.7 and the parameters mentioned below. Since the CHs are elected in a random manner, we execute simulation for each set of parameters independently 100 times. The results shown in this section are the average results.

The parameters used in this section are summarized in Table 5.2. The parameters for the energy consumption model are consistent with [10,15].

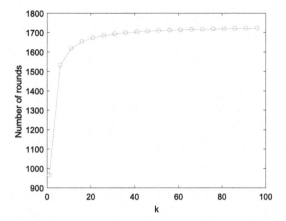

FIGURE 5.8 The impact of k on the number of rounds.

5.5.1 Parameter impacts

There are several parameters in our protocol, namely α, β, γ, k, and R_0. Here β is the threshold for whether a node can act as CH candidate. The higher the value of β, the smaller the possibility for a node to become a CH candidate. We fix β as 0.1. The factor γ is used to adjust the communication range for a CH in CH-CH association. The higher the value of γ, the more neighbor CHs can be found. However, large values for γ lead to energy waste, since energy consumption is a function of transmission distance. Also, when the value of γ is too high, all CHs may be associated to one CH with a low hop number and having the highest value for η, which heavily increases the relay burden of this CH. We fix γ as 2. As mentioned in Section 5.3.2, the higher the value of k, the fewer control messages need to be transmitted. But too high k values may cause some CHs to die within a cycle. We consider the influence of k on the lifetime of a cluster. Here, we select one cluster and test the number of rounds it can operate with the parameters in Table 5.2. As shown in Fig. 5.8, the lifetime increases with k because larger k means more energy is spent on data transmission and less on overhead. We can also see that when k is smaller than 20, the lifetime raises quickly; after that, it increases slightly. Thus, we fix k as 30 since it gives a relatively long lifetime.

Next, we study how α and R_0 influence the protocol performance, with α ranging from 0.2 to 0.6 and R_0 from 10 to 50. We consider the number of clusters under different parameter sets in both Scenario 1 and Scenario 2. The result is shown in Fig. 5.9a. For a fixed α, the number of clusters decreases with increasing R_0. According to (5.5), we know that when α is fixed, the cover range increases with R_0. It testifies our design purpose, i.e., the larger the cover range, the smaller the number of clusters. Besides, for a fixed R_0, the number of clusters increases with increasing α. The reason behind this is as follows. According to (5.5), given R_0, α influences the range of R_i. The higher the value of α, the

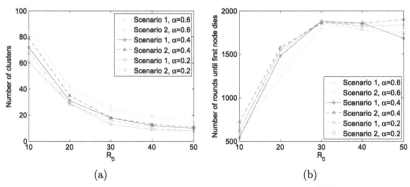

FIGURE 5.9 The impacts of α and R_0 on (a) the number of clusters and (b) the number of rounds until the first node dies.

smaller the lower bound of R_i. In other words, R_i can take more values when α is set as a larger value. Thus, the number of clusters also increases with α.

Now we consider the network lifetime under different parameter sets. The network lifetime is significant in many applications since the sensing field cannot be fully monitored once a node dies. The result is shown in Fig. 5.9b. For a fixed α, with increasing R_0, the network lifetime tends to increase. The reason behind this is as follows. As we assume that each CH aggregates the data packets received from its CMs into one packet, the larger the number of clusters, the more data packets are transmitted, resulting in a larger amount of energy consumption. This analysis is consistent with the result shown in Fig. 5.9b, i.e., the network lifetime increases with increasing R_0.

5.5.2 Comparison with existing work

Since the advantage of applying MS to collect data in WSNs has been shown in the literature, to make a fair comparison we compare our approach with those which consider the scenario of using a path-constrained MS. Here, we compare our results with those reported in [17] and [18], where two representative approaches for data collection based on flat and cluster structures, respectively, were used. The approach in [18] determines the cover range for each CH according to the Euclidean distance to the MS's trajectory. Thus, it requires the sensor nodes to be able to extract the distance information from the received signal strength. Furthermore, it also requires all the sensor nodes to be within the range of the MS. Thus, considering the size of the sensing field, the MS communication range is set to be 100 m. In contrast, since our approach and [17] use only hop distance, the communication range of the MS is set as 30 m. Since the approach in [18] constructs the clusters with two sizes and considering the sensing field, the communication range of sensor nodes is set as 60 m. The communication ranges of sensor nodes for the approach in [17] and ours are set

TABLE 5.3 Comparison of approaches.

Approach	MASP [17]	MobiCluster [18]	Ours
Distance measure	No	Yes	No
Network structure	Flat	Cluster	Cluster
MS comm. range	30 m	100 m	30 m
Node comm. range	30 m	60 m	30 m

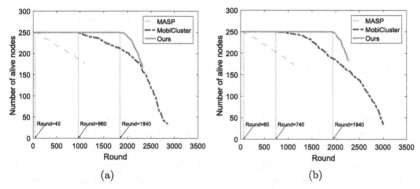

(a)   (b)

FIGURE 5.10 Number of alive nodes in each scenario. (a) Uniform node distribution. (b) Nonuniform node distribution.

at the same value as that of the MS. The basic features of our approach, [17], and [18] are shown in Table 5.3.

Here our performance measure is still the network lifetime. The results of our approach and the other two are shown in Fig. 5.10. In both Scenario 1 and Scenario 2, the proposed approach achieves the longest network lifetime, which guarantees the network to have good coverage of the interested areas for a long time. Besides, with the same initial amount of energy, no matter if the nodes are uniformly distributed or nonuniformly distributed, the networks can achieve a similar network lifetime using the proposed approach. However, the approach of [18] performs differently, i.e., in the uniform node distribution case, it can operate for 960 rounds with all the nodes alive, while in the nonuniform node distribution case, it can only perform 740 rounds. The reason is that it does not take node density into account. Since the approach of [17] is based on a flat structure, the node distribution impacts little. It performs similarly in both scenarios and exhibits the worst performance in both, because the subsinks suffer from funneling effect issues. Comparing the performance of our approach and [18], it can be seen that after the first node dies in these two cases, the number of alive nodes in our protocol drops more dramatically than in [18]. This is because in our protocol the energy consumption of the sensor nodes is more balanced than that of [18]. Fig. 5.10 also plots for how many rounds each protocol

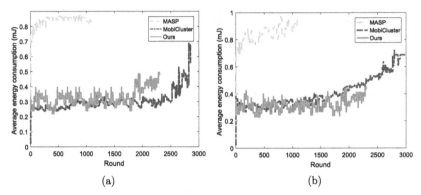

FIGURE 5.11 Average energy consumption by the alive nodes. (a) Uniform node distribution. (b) Nonuniform node distribution.

can operate under the connectivity requirement.[2] In this context, the approach of [18] results in the longest operation, because the sensor nodes' communication range is larger. Our approach and [17] result in shorter operations due to relatively short communication ranges. Although it is possible to increase the number of operation rounds by increasing the communication range, it is not necessary since the network has already lost full coverage of the sensing field.

Accordingly, the average energy consumptions by the alive nodes in these two scenarios are displayed in Fig. 5.11. The values on round 0 represent the average energy consumption in the initial phase while later values represent the collection phase. The energy consumption at the beginning of each cycle, i.e., for clustering, is added to the first round. In the initial phase, the proposed approach consumes more energy than the alternatives, since it requires three rounds of control message exchange. In the collection phase, the average energy consumption increases when some nodes run out of energy. The fundamental reason is that the average transmission distance increases when some nodes die. Comparing Figs. 5.11a and 5.11b, we can see that the proposed approach results in lower energy consumption than those of [17,18] in the case of a nonuniform distribution.

We further consider four other networks with 200 and 300 nodes which are uniformly and nonuniformly deployed, respectively. We also apply the approaches compared here on these networks. As shown in Fig. 5.12, the network lifetimes with our proposed protocol are the longest.

Moreover, we provide a simulation based on the real trace of the shuttle in Wollongong, Australia. The shuttle route is shown in Fig. 5.13a and the timetable can be found on the website of Transport for NSW. One hundred sensor nodes are nonuniformly deployed in this 3 km × 5 km area (the right part of the area has a higher density than the left part). We apply the proposed approach

---

[2] Connectivity requirement: The sensory data packet at each alive sensor node can be transmitted to the MS. In other words, every alive sensor node should have at least one neighbor node.

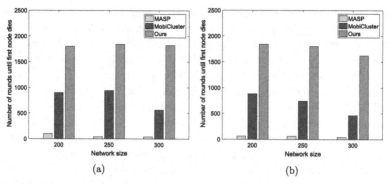

FIGURE 5.12 Network lifetime over network sizes. (a) Uniform node distribution. (b) Nonuniform node distribution.

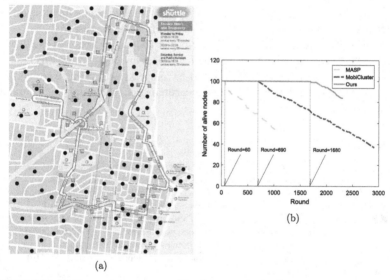

FIGURE 5.13 (a) The sensing area covered by the route of a shuttle and a set of sensor nodes. (b) Number of alive nodes.

as well as the two alternatives. For this practical case, buffer overflow may occur since the shuttle does not operate at night. Here we do not consider the buffer overflow issue and only consider the network lifetime. The simulation results are shown in Fig. 5.13b. We can see that the proposed approach achieves the longest network lifetime.

5.6 Summary

We propose a cluster-based routing protocol to support data collection in WSNs with nonuniform node distribution using a path-fixed MS. It involves an energy-aware and unequal clustering algorithm and an energy-aware routing algorithm. The clustering algorithm constructs clusters with unequal ranges based on the local node density and the hop distance to the MS's trajectory. The resulting clusters have the following features: (1) at the same hop distance, the clusters in the dense (sparse) area have small (large) cover ranges; and (2) with the same density, the clusters close to (far away from) the MS's trajectory have small (large) cover ranges. Such formation helps balance the energy consumption among clusters. In the routing algorithm, each CH selects a CH with a lower hop distance to the MS's trajectory and the highest residual energy/CM size as the relay CH. By using the above techniques, our protocol works well for WSNs with nonuniform node distribution and prolongs the network lifetime significantly compared to existing work. Moreover, since the Euclidean distance measure is not required, the system cost is low, which gives the proposed protocol good scalability properties. The proposed approach is for a single MS. When more MSs are available, the cooperation between them may help to increase the performance further, which will be the subject of our future research.

In this chapter, we study the cluster-based routing protocol assisted by a single MS. Local aggregation is assumed at CHs. Obviously, using this method some information in the original data is lost. Compressive sensing, which is able to recover the original data from a few measurements, is a promising technique to address this concern.

References

[1] J. Li, P. Mohapatra, Analytical modeling and mitigation techniques for the energy hole problem in sensor networks, Pervasive and Mobile Computing 3 (3) (2007) 233–254.
[2] M. Di Francesco, S.K. Das, G. Anastasi, Data collection in wireless sensor networks with mobile elements: a survey, ACM Transactions on Sensor Networks 8 (1) (2011) 7.
[3] Y. Gu, F. Ren, Y. Ji, J. Li, The evolution of sink mobility management in wireless sensor networks: a survey, IEEE Communications Surveys and Tutorials 18 (1) (2015) 507–524.
[4] M. Zhao, Y. Yang, Optimization-based distributed algorithms for mobile data gathering in wireless sensor networks, IEEE Transactions on Mobile Computing 11 (10) (2012) 1464–1477.
[5] S. Mottaghi, M.R. Zahabi, Optimizing leach clustering algorithm with mobile sink and rendezvous nodes, AEÜ. International Journal of Electronics and Communications 69 (2) (2015) 507–514.
[6] T.M. Cheng, A.V. Savkin, Decentralized control of mobile sensor networks for asymptotically optimal blanket coverage between two boundaries, IEEE Transactions on Industrial Informatics 9 (1) (2013) 365–376.
[7] A.V. Savkin, F. Javed, A.S. Matveev, Optimal distributed blanket coverage self-deployment of mobile wireless sensor networks, IEEE Communications Letters 16 (6) (2012) 949–951.
[8] A.V. Savkin, T.M. Cheng, Z. Xi, F. Javed, A.S. Matveev, H. Nguyen, Decentralized Coverage Control Problems for Mobile Robotic Sensor and Actuator Networks, John Wiley & Sons, 2015.
[9] H. Huang, A.V. Savkin, An energy efficient approach for data collection in wireless sensor networks using public transportation vehicles, AEÜ. International Journal of Electronics and Communications 75 (2017) 108–118.
[10] G. Chen, C. Li, M. Ye, J. Wu, An unequal cluster-based routing protocol in wireless sensor networks, Wireless Networks 15 (2) (2009) 193–207.

[11] D. Wei, Y. Jin, S. Vural, K. Moessner, R. Tafazolli, An energy-efficient clustering solution for wireless sensor networks, IEEE Transactions on Wireless Communications 10 (11) (2011) 3973–3983.

[12] X. Min, S. Wei-Ren, J. Chang-Jiang, Z. Ying, Energy efficient clustering algorithm for maximizing lifetime of wireless sensor networks, AEÜ. International Journal of Electronics and Communications 64 (4) (2010) 289–298.

[13] J. Yu, Y. Qi, G. Wang, X. Gu, A cluster-based routing protocol for wireless sensor networks with nonuniform node distribution, AEÜ. International Journal of Electronics and Communications 66 (1) (2012) 54–61.

[14] H. Li, Y. Liu, W. Chen, W. Jia, B. Li, J. Xiong, Coca: constructing optimal clustering architecture to maximize sensor network lifetime, Computer Communications 36 (3) (2013) 256–268.

[15] M. Shokouhifar, A. Jalali, A new evolutionary based application specific routing protocol for clustered wireless sensor networks, AEÜ. International Journal of Electronics and Communications 69 (1) (2015) 432–441.

[16] W.R. Heinzelman, A. Chandrakasan, H. Balakrishnan, Energy-efficient communication protocol for wireless microsensor networks, in: Proceedings of the 33rd Annual Hawaii International Conference on System Sciences, IEEE, 2000, pp. 1–10.

[17] S. Gao, H. Zhang, S.K. Das, Efficient data collection in wireless sensor networks with path-constrained mobile sinks, IEEE Transactions on Mobile Computing 10 (4) (2011) 592–608.

[18] C. Konstantopoulos, G. Pantziou, D. Gavalas, A. Mpitziopoulos, B. Mamalis, A rendezvous-based approach enabling energy-efficient sensory data collection with mobile sinks, IEEE Transactions on Parallel and Distributed Systems 23 (5) (2012) 809–817.

Chapter 6

Energy-efficient path planning of a solar-powered UAV for secure communication in the presence of eavesdroppers and no-fly zones[☆]

6.1 Motivation

With the rapidly growing demand for energy and decreasing fossil fuel reserves, renewable energy has attracted much attention worldwide in the last few decades. Solar power is an important source of renewable energy, and various solar-powered products have become part of our daily life, such as solar-powered cars [1] and solar-powered UAVs [2]. Powering UAVs with solar energy is a promising solution to overcome the bottleneck of the limited flight time of UAVs powered by onboard batteries. Harvesting solar energy while flying allows UAVs to fly longer in the sky, which enables the UAVs to provide uninterrupted service in different applications such as monitoring and surveillance [3,4], disaster management [5], infrastructure inspection [6,7], parcel delivery [8], and wireless communication [9].

The application of UAVs in wireless communication is a hot topic. Compared to conventional wireless communication systems based on stationary infrastructures, UAV-based communication systems can provide timely and economical services in temporary hotspots, disaster areas, and complex terrains [9,10], thanks to their maneuverability. A good number of research articles relevant to UAV-enabled wireless communication systems have been published, and the UAVs assist existing communication systems by offloading the traffic demand of the stationary access points [9,11,12] and acting as relays to connect remote users to networks [13].

UAVs have also been used to protect wireless communication systems. Paper [14] maximizes the nonzero secrecy capacity of a UAV–node link in the presence of an eavesdropper by optimally deploying the UAV. Reference [15]

[☆] The main results of the chapter were originally published in Hailong Huang, Andrey V. Savkin, Wei Ni, Energy-efficient 3D navigation of a solar-powered UAV for secure communication in the presence of eavesdroppers and no-fly zones, Energies 13 (6) (2020) 1445.

https://doi.org/10.1016/B978-0-32-390182-6.00011-2
Copyright © 2022 Elsevier Inc. All rights reserved.

considers a four-entity system with a transmitter, a receiver, an eavesdropper, and a UAV relay. It optimizes the transmit power of the transmitter and the UAV to maximize the secrecy rate of the system. Paper [16] uses a UAV as a jammer that interferes with eavesdroppers. The paper focuses on optimizing the trajectory and the jamming power of the UAV and the transmit power of the transmitter to maximize the average achievable secrecy rate. Reference [17] studies a protection scheme using two UAVs: one transmits information, and the other interferes with eavesdroppers. The study jointly optimizes the UAVs' trajectories and the user transmission schedule. Paper [18] investigates a scenario where two UAVs transmit confidential messages to specified ground nodes using the same spectrum in the presence of an eavesdropper. Though the mutual cochannel interferences arising from spectrum sharing may lower the spectral efficiency, the eavesdropping effect can be reduced as the two UAVs act in essence as cooperative jammers for each other. Papers [19,20] secure the wireless data communication between a ground transmitter and a UAV by jointly optimizing the trajectory of the UAV and the transmit power of the ground transmitter to maximize the average secrecy rate over a given period of time. One common setting of [14–20] is that the UAV is powered by an onboard capacity-limited battery. Therefore, these designs may not ensure a sustainable defense against eavesdropping.

This chapter considers using a solar-powered UAV to secure wireless communication systems with a ground node in the presence of collaborative eavesdroppers. By securing wireless communication, we mean that the ground node can recover the data sent by the UAV and the eavesdroppers cannot do it at any time. This is a new scenario that has not been considered comprehensively in the literature. Firstly, in light of [21], we consider a solar-powered UAV which is possible to operate for a longer time than the onboard battery-powered UAVs. Secondly, unlike [16,18–20] which optimize the integrated jamming or eavesdropping performance, we target on instantaneous communication protection. Thirdly, in light of [22], the eavesdroppers can collaborate by combining the collected signals to get a better understanding of the information sent by the UAV. Obviously, this can make it more difficult for the UAV to optimize its trajectory. Moreover, different from most existing publications which assume that the UAV flies in free space, we consider no-fly zones. The no-fly zones may be tall buildings or areas with high risks of being targeted by ground-to-air missiles [23–25]. These different considerations make the problem considered in this chapter more general, challenging, and practical.

In this chapter, we focus on a new trajectory optimization problem for a UAV by taking into account solar power harvesting, instantaneous communication protection, collaborative eavesdropping, and no-fly zones. We propose a new optimization model to minimize the energy expenditure of a UAV with a solar-powered rechargeable battery while securing communications between the UAV and a ground node, avoiding no-fly zones, satisfying the aeronautic maneuverability, and preventing depletion of the battery. This is a realistic problem but needs a nonconvex optimization technique, as the solar power is nonsmooth and the UAV dynamic model is nonlinear. We propose a scheme to plan the UAV trajectory based on RRT. In the proposed method, an RRT can be constructed to capture the nonlinear UAV motion model and the secure communication requirements. From the RRT, we can randomly select a set of possible trajectories which end in a predefined destination set. (The final position of the UAV is in the destination set.) We verify the energy storage along each possible trajectory and choose the one with the minimum energy expenditure while satisfying all the constraints. Our key contributions include the new problem formulation and the new RRT-based trajectory planning method, which is computationally efficient and can explore the flight space to construct UAV trajectories fast. Computer simulations demonstrate the effectiveness of the proposed approach. By comparing to a baseline method that is unaware of eavesdropping, the proposed method is shown to guarantee the valid wireless communication link with the ground node and prevent eavesdropping. Though a static environment setting is demonstrated in the chapter, the proposed RRT-based method can be potentially applied to dynamic environments thanks to its computational efficiency.

RRT falls into the category of sampling-based path planning algorithms. A similar algorithm is a probabilistic roadmap (PRM). It first builds up a dense enough roadmap (graph) and finds a path from the start to the destination by a graph search algorithm [26]. By contrast, the RRT method achieves the best feasible path to the goal all by its own processing procedure, which is more suitable for dynamic cases. Another category of path planning methods is the node-based algorithms such as Dijkstra's algorithm [27]. They grid the solution space and need a specific metric to characterize the cost of moving from one grid to another. These algorithms cannot handle the mobility constraints of UAVs such as the nonholonomic constraints, as they generally regard the robot as a point in the graph [27]. However, the RRT-based method can account for the mobility constraint in the process of constructing the tree. Furthermore, as will be shown later, it is not straightforward to define the cost of moving from one grid to another for the node-based algorithms. From these discussions, the RRT-based method well suits the considered problem. The main results of the chapter were originally published in [28].

The remainder of the chapter is organized as follows. In Section 6.2, we present the system models and state the problem under investigation. In Section 6.3, we present the proposed 3D trajectory optimization method to minimize the energy consumption subject to the constraints of the UAV maneuverability, communication security, battery lifetime, and no-fly zones. Computer simulations are conducted in Section 6.4 to show the performance of the proposed navigation algorithm. Finally, a summary is given in Section 6.5.

6.2 Problem statement

In this section, we first present the system model and then formulate the problem of interest.

6.2.1 UAV dynamics

We consider a UAV flying in a 3D space. Let $p(t) = [x(t), y(t), z(t)]$ be the coordinates of the UAV at time t, where $x(t)$ and $y(t)$ are the coordinates on the horizontal plane parallel to the ground and $z(t)$ is the altitude of the UAV. We consider the following well-known model of UAV motion[1]:

$$\begin{cases} \dot{x}(t) = v(t)\cos(\theta(t)), \\ \dot{y}(t) = v(t)\sin(\theta(t)), \\ \dot{\theta}(t) = \omega(t), \\ \dot{z}(t) = u(t), \\ 0 \le v(t) \le V^{max}, \\ -\Omega^{max} \le \omega(t) \le \Omega^{max}, \\ -U^{max} \le u(t) \le U^{max}, \end{cases} \tag{6.1}$$

where $\theta(t)$ is the heading of the UAV with respect to the x-axis; $v(t)$, $\omega(t)$, and $u(t)$ are its linear horizontal, angular, and vertical speeds, respectively, and V^{max}, Ω^{max}, and U^{max} are given positive constants indicating the maximum linear, angular, and vertical speeds, respectively. The linear horizontal speed $v(t)$, angular speed $\omega(t)$, and vertical speed $u(t)$ are considered as the control inputs of the model in (6.1). The UAV altitude $z(t)$ should satisfy the following constraint:

$$Z^{min} \le z(t) \le Z^{max}, \tag{6.2}$$

where $Z^{min} > 0$ is the lower bound of the UAV altitude and $Z^{max} > 0$ is the upper bound. Furthermore, the horizontal position of the UAV must be outside no-fly zones $\mathcal{Z}$:

$$(x(t), y(t)) \notin \mathcal{Z}. \tag{6.3}$$

We assume that the set of no-fly zones $\mathcal{Z}$ consists of r no-fly zones $\mathcal{Z}_1, \ldots, \mathcal{Z}_r$, where $\mathcal{Z}_1, \ldots, \mathcal{Z}_r$ are bounded and connected planar sets. These no-fly zones may be urban areas where flight of UAVs is prohibited. Or, in military applications the no-fly zones may represent areas with high risks of being targeted by ground-to-air missiles [23–25].

6.2.2 Harvesting solar energy

The UAV can harvest solar energy, and the harvesting rate depends on the cloud condition if there is any. We consider cloudy weather, and there is a cloud layer with an altitude ranging from H_l to H_u. The output power of the solar panel

---

[1] This 3D UAV motion model is extended from a 2D model [29] where a forward speed and an angular speed are the control inputs. Beyond these two control inputs, we consider the vertical speed such that the UAV can adjust its altitude.

$(P_{solar}(t))$ can be modeled by a function of the altitude $(z(t))$ [30]:

$$P_{solar}(t) = \begin{cases} \eta SG, z(t) \geq H_u, \\ \eta SGe^{-\beta(H_u-z(t))}, H_l \leq z(t) \leq H_u, \\ \eta SGe^{-\beta(H_u-H_l)}, z(t) < H_l, \end{cases} \tag{6.4}$$

where η is a constant representing the energy transfer efficiency of solar energy harvesting, S is the area of the solar panels, G is the average solar radiation intensity on the earth, and β is the absorption coefficient of the solar panels. From (6.4), we see that when the UAV is above the cloud, the solar energy harvesting power $P_{solar}(t)$ is the largest; when it is below the cloud, $P_{solar}(t)$ is the lowest; and when it is inside the cloud, $P_{solar}(t)$ increases exponentially with the altitude. Eq. (6.4) is a piecewise function of $z(t)$. The nonsmoothness makes the problem under consideration complex to solve.

6.2.3 UAV energy consumption

The energy consumption of the UAV depends on the velocity vector $v(t) = (\dot{x}(t), \dot{y}(t), \dot{z}(t))$ [21], and can be modeled as the summation of the induced power for level flight, the power for vertical flight, and the profile power related to the blade drag.

The induced power for level flight is modeled as

$$P_{lf}(t) = \frac{W^2}{\sqrt{2}\rho A} \frac{1}{\sqrt{\|(\dot{x}(t), \dot{y}(t))\|^2 + \sqrt{\|(\dot{x}(t), \dot{y}(t))\|^4 + 4V_h^4}}}, \tag{6.5}$$

where W is the UAV weight, ρ is the air density, A is the total area of the UAV rotor disks, and $V_h = \sqrt{\frac{W}{2\rho A}}$.

The power for vertical flight is modeled as

$$P_{vf}(t) = W\dot{z}(t). \tag{6.6}$$

The profile power related to the blade drag is modeled as

$$P_{bg}(t) = \frac{1}{8}C_{D0}\rho A\|(\dot{x}(t), \dot{y}(t))\|^3, \tag{6.7}$$

where C_{D0} is the profile drag coefficient depending on the geometry of the rotor blades.

Thus, at any time t, the power consumption of the UAV for movement is:

$$P_{UAV}(t) = P_{lf}(t) + P_{vf}(t) + P_{bg}(t). \tag{6.8}$$

Moreover, let P_0 be the static power consumed for maintaining the operation of the UAV and let P_t be the constant transmit power. Let $q(t)$ denote the energy stored in the onboard battery. Then, the dynamics of the stored energy of the battery are characterized by

$$\dot{q}(t) = P_{solar}(t) - P_{UAV}(t) - P_0 - P_t. \tag{6.9}$$

Let q_c denote the battery capacity. Then, the following constraint should be satisfied:

$$q(t) \leq q_c. \tag{6.10}$$

6.2.4 Securing communication

We consider that the UAV communicates with a stationary ground node $\mathcal{N}$. Let (x_n, y_n) be the position of the node. Moreover, there are M stationary eavesdroppers on the ground, which attempt to eavesdrop on the communication between the UAV and the node $\mathcal{N}$. Let (x_i^e, y_i^e) be the position of an eavesdropper i, $i = 1, \ldots, M$.

Let $d(t)$ denote the Euclidean distance between the UAV and the node $\mathcal{N}$ at time t, and let $d_i(t)$ be the Euclidean distance between the UAV and the eavesdropper i at time t:

$$d(t) = \sqrt{(x(t) - x_n)^2 + (y(t) - y_n)^2 + z(t)^2},$$
$$d_i(t) = \sqrt{(x(t) - x_i^e)^2 + (y(t) - y_i^e)^2 + z(t)^2}.$$

We introduce a function $f(d)$ which measures the communication/eavesdropping performance from the distance d. Clearly, $f(d)$ is a decreasing function of d. For example, in the well-known Friis formula [31], when the transmitter and receiver are d apart with dominating LoS, the received SNR is $f(d) = \frac{\phi}{d^2}$, where $\phi > 0$ is a given parameter depending on the transmit power, the gains of the transmitter and receiver, and the noise power.

To guarantee secure communication between the UAV and the ground node, a set of constraints are considered below. The first requirement is that the received power at the ground node needs to be no lower than a threshold γ_c:

$$f(d(t)) \geq \gamma_c, \ \forall t \geq 0. \tag{6.11}$$

This is to ensure that the node can successfully recover the received data. Secondly, the received power at each eavesdropper needs to be no larger than a threshold γ_e:

$$f(d_i(t)) \leq \gamma_e, \ \forall t \geq 0, \tag{6.12}$$

for all $i = 1, \ldots, M$. This is to ensure that any eavesdropper i cannot recover the captured data. Lastly, the following constraint is satisfied to avoid eavesdropping when the eavesdroppers collaborate:

$$\sum_{i=1}^{m} f(d_i(t)) \leq \gamma_{ce}, \ \forall t \geq 0, \tag{6.13}$$

where $\gamma_{ce} \leq m\gamma_e$. The constraint (6.13) ensures that when the eavesdroppers collaborate, they cannot recover the combined data. When they carry out maximal ratio combining (MRC) [31], (6.13) can assume equality. General speaking, the constraints (6.12) and (6.13) require the UAV to be far enough from the eavesdroppers, whereas (6.11) expects the UAV to be close to the node $\mathcal{N}$. We note that constraints (6.11), (6.12), and (6.13) specify stringent instantaneous security rate requirements. We also note that in cases where several eavesdroppers are located closely, they may block the path for the UAV, and there may not exist a feasible trajectory satisfying (6.11), (6.12), and (6.13). To avoid being eavesdropped, we assume that the UAV stops transmitting in these cases.

6.2.5 Problem statement

Let $T > 0$ be the time the UAV can use to fly to a destination set $\mathcal{D}$. We assume that the destination set $\mathcal{D}$ is a bounded and connected planar set that does not overlap with the set of no-fly zones $\mathcal{Z}$. During the period $[t_0, t_0 + T]$, the UAV keeps transmitting confidential data to the stationary ground node. The problem of interest is formulated as follows:

$$\min_{v(t),\omega(t),u(t)} \int_{t_0}^{t_0+T} P_{UAV}(t) dt \tag{6.14}$$

subject to

$$0 < q(t) \leq q_c, \tag{6.15}$$

$$(x(t_0 + T), y(t_0 + T)) \in \mathcal{D}, \tag{6.16}$$

$$(6.2), (6.3), (6.11), (6.12), \text{ and } (6.13).$$

In other words, this problem is to develop a navigation law for the UAV (6.1) that minimizes (6.14) subject to the constraints (6.2), (6.3), (6.11), (6.12), (6.13), (6.15), and (6.16).

6.3 Navigation law

The problem under consideration cannot be solved using convex optimization tools, because of the nonconvexity of the objective function (6.14), especially the part for level flight (6.5), the nonsmoothness of the solar power harvesting model (6.4), and the nonlinearity of the UAV dynamic model (6.1). We propose an RRT-based trajectory planning scheme to solve the problem heuristically and efficiently. The RRT method can tackle the nonlinear dynamic model, secure communication requirements and no-fly zones by generating random samples in the searching space, and only keeping the samples satisfying all the constraints.

We assume that the UAV knows the locations of the node and the eavesdroppers. We divide $[t_0, t_0 + T]$ into L equal intervals (each interval lasts $\delta = \frac{T}{L}$) and assume that the control inputs $v(t)$, $\omega(t)$, and $u(t)$ are updated at time instants $t_0, t_0 + \delta, \ldots, t_0 + (L-1)\delta$ and remain constant at any interval $[t_0 + j\delta, t_0 + (j+1)\delta]$, $j = 0, 1, \ldots, L-1$. The control inputs of the UAV dynamic model are simplified as follows:

$$\begin{cases} w(j) = V^{max}, \\ w(j) \in \{-\Omega^{max}, 0, \Omega^{max}\}, \\ u(j) \in \{-U^{max}, 0, U^{max}\}, \end{cases} \tag{6.17}$$

for $j = 0, 1, \ldots, L - 1$. Compared to (6.1), the input $v(j)$ keeps the maximum value V^{max}, and $\omega(j)$ and $u(j)$ in (6.17) either take 0 or the corresponding maximum value, respectively. Though this simplification sacrifices the maneuverability of the UAV, it is still realizable since the selected control inputs are all within the allowed ranges. More importantly, this simplification makes the UAV easy to control and also narrows down the solution space significantly. It is not hard to see that this method can be easily extended to include some more feasible control inputs.

Based on (6.17), the models (6.4), (6.11), (6.12), and (6.13) can be discretized directly as they are functions of the UAV position $p(j)$. To discretize the energy consumption model of the UAV, we rewrite the velocity vector $v(j)$ as $(\frac{x(j+1)-x(j)}{\delta}, \frac{y(j+1)-y(j)}{\delta}, \frac{z(j+1)-z(j)}{\delta})$. Then, the discrete forms of (6.8) and (6.14) can be obtained. Accordingly, the relationship of (6.9) is rewritten as

$$q(j + 1) = q(j) + \delta(P_{solor}(j) - P_{UAV}(j) - P_0 - P_t). \tag{6.18}$$

Considering the battery capacity q_c, (6.18) should be corrected by

$$q(j + 1) = \max\{q(j), q_c\}. \tag{6.19}$$

The trajectory of the UAV for the interval $[t_0, t_0 + T]$ can be found by solving the discrete version of the considered problem. Specifically, having the position and heading of the UAV at t_0, we can construct an RRT with a certain number of vertices based on (6.17), (6.2), (6.3), (6.11), (6.12), and (6.13). Then, any vertex of the RRT is reachable from the root via a sequence of control inputs defined in (6.17), is within the allowed altitudes, is outside the no-fly zones, and can avoid both individual and collaborative eavesdropping. However, not all the vertices in the RRT are feasible for the UAV to visit due to the constraints (6.15) and (6.16). From the RRT, we can randomly select a trajectory with L consecutive vertices starting from the root. If the last vertex falls into the destination set $\mathcal{D}$, then the constraint (6.16) holds. Furthermore, for each vertex of this trajectory, we verify the constraint (6.15). If (6.15) also holds, we finally compute the energy consumption of the UAV along this trajectory according to (6.14). We can randomly select a set of trajectories as above, and the one with the minimum energy consumption can be identified. This algorithm is summarized as follows.

Algorithm:

1. Construct an RRT rooted at the initial position of the UAV with a certain number of vertices falling into the destination set $\mathcal{D}$ and all the vertices satisfy (6.17), (6.2), (6.3), (6.11), (6.12), and (6.13).
2. Randomly select a certain number of trajectories with no more than N consecutive vertices from the RRT which start from the root and end at a vertex belonging to $\mathcal{D}$.
3. For each of the selected trajectories, verify whether constraints (6.15) and (6.16) are satisfied.
4. Among the trajectories satisfying (6.15) and (6.16), identify the one with the minimum energy consumption according to (6.14).

The RRT-based method can explore the solution space fast. It belongs to the group of sampling-based algorithms. Another sampling-based method is the PRM. This method builds up a roadmap (graph) first which should be dense enough such that a path exists between the start and the destination. Then, a path can be obtained by the shortest path query. Different from the RRT-based method, which achieves the best feasible path to the goal all by its own processing procedure, PRM generally calls a graph searching algorithm [26]. Another group of path planning methods is the node-based algorithms. Dijkstra's algorithm and its variants are the typical ones. Being node-based, they require to grid the solution space, and a specific metric is needed to characterize the cost of moving from one grid to another [27]. These algorithms cannot handle the mobility constraints of UAVs such as the nonholonomic constraints. Specifically, in the gridded space, moving from one node to another may be impossible for a UAV having limited angular speed. If possible, the computation of the feasible linear, angular, and vertical speeds is required. However, the developed RRT-based method takes the mobility constraint in the process of constructing the tree into account. Thus, the obtained trajectory can be tracked directly without any later adjustment. Furthermore, for the problem under consideration, it is not straightforward to define the cost of moving from one grid to another for the node-based algorithms. From these discussions, the RRT-based method well suits the considered problem.

6.4 Simulation results

In this section, we demonstrate the effectiveness of the proposed method via computer simulations. The parameters used are summarized in Table 6.1. We consider two simulation environments, as shown in Figs. 6.1 and 6.2. In Fig. 6.1, there are two eavesdroppers and four no-fly zones, while in Fig. 6.2, there are

TABLE 6.1 Parameters and values.

Parameter	Value	Parameter	Value
S	1 m^2	G	1367 W/m^2
η	0.6	β	0.01
H_l	700 m	H_u	1400 m
ρ	1.225 kg/m^3	A	0.5 m^2
W	18.4 N	C_{D0}	0.08
P_0	5 W	$q(0)$	100 wh
q_c	150 wh	V^{max}	5 m/s
Ω^{max}	0.5 rad/s	U^{max}	0.5 m/s
Z^{min}	100 m	Z^{max}	1600 m
ϕ	1.8×10^6	γ_c	2
γ_e	1.5	γ_{ce}	1.6
δ	1 minute	T	30 minutes

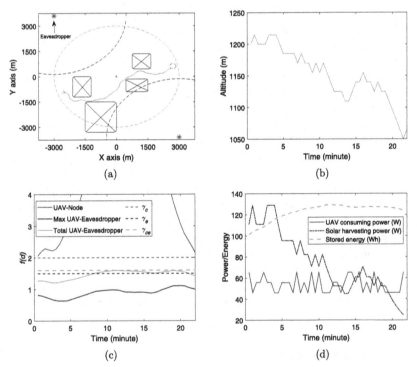

FIGURE 6.1 Simulation results in Environment 1 by the proposed method. (a) The UAV movement on the xy-plane. A video is available at https://youtu.be/7EpmQZuOwao. (b) The UAV altitude. (c) The communication and eavesdropping performance. (d) Energy consumption and solar energy harvesting.

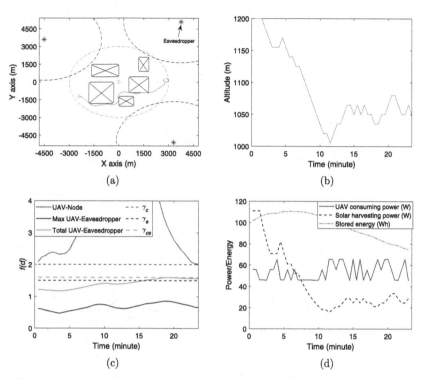

FIGURE 6.2 Simulation results in Environment 2 by the proposed method. (a) The UAV move-ment on the xy-plane. A video is available at https://youtu.be/BiNL8gA0n-w. (b) The UAV altitude. (c) The communication and eavesdropping performance. (d) Energy consumption and solar energy harvesting.

three eavesdroppers and five no-fly zones. The ground node is located at $(0, 0)$. The UAV starts from the initial position, which is shown by the green trian-gle in Figs. 6.1a and 6.2a. The destination set is shown by a blue circle on the right side of Figs. 6.1a and 6.2a. Figs. 6.1a and 6.2a, we also demonstrate some dash lines on the plane of $z = 0$, i.e., the ground. The purple (mid gray in print version) one indicates the area within which the ground node can ef-fectively recover the data sent by the UAV. The black dash lines (dark gray in print version) represent the area within which the eavesdroppers (marked by the black squares) can recover the captured data. Starting from the initial position, the UAV moves along the red trajectory and reaches the destination set in 22 minutes in Environment 1 and in 23.5 minutes in Environment 2, which are be-fore the due time. The altitude of the UAV during the movement is shown in Figs. 6.1b and 6.2b. The communication between the UAV and the ground node is good, as seen in Figs. 6.1c and 6.2c, and this trajectory avoids the individ-ual and collaborative eavesdropping (Figs. 6.1c and 6.2c). Finally, the energy consumption of the UAV, the solar energy harvesting, and the stored amount of

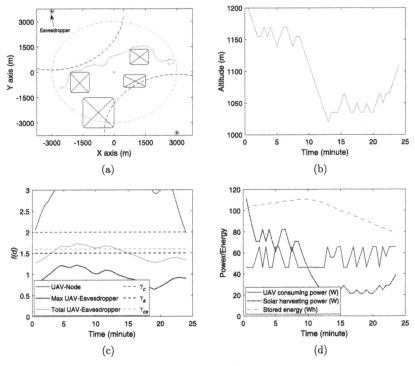

FIGURE 6.3 Simulation results in Environment 1 by the benchmark method. (a) The UAV movement on the xy-plane. A video is available at https://youtu.be/kGFlpWV-SkQ. (b) The UAV altitude. (c) The communication and eavesdropping performance. (d) Energy consumption and solar energy harvesting.

energy are shown in Figs. 6.1d and 6.2d. Note that the simulations are conducted on a normal PC with Intel(R) Core(TM) i7-8565U CPU @ 1.8 GHz, 1.99 GHz and 8G RAM. The average time to find the trajectory is about 3–5 seconds. Specifically, for the simulation in Environment 1, the algorithm generates about 800–1000 samples. The number of samples generated changes in each simulation as the RRT-based algorithm is a random method. For Environment 2, the algorithm generates more samples, i.e., 10,000–13,000. The reason is that Environment 2 (three eavesdroppers and five no-fly zones) is more complex than Environment 1 (two eavesdroppers and four no-fly zones). Thus, there are more infeasible samples generated in Environment 2.

For comparison, we apply a baseline method to the considered case. This baseline method only accounts for effectively transmitting data to the ground node while ignoring the eavesdroppers. All the parameters are the same as above. The simulation results are shown in Figs. 6.3 and 6.4. The trajectories of the UAV are shown in Figs. 6.3a and 6.4a. Comparing Fig. 6.3a and Fig. 6.1a (Fig. 6.4a and Fig. 6.2a) we see that this trajectory is closer to the eavesdropping

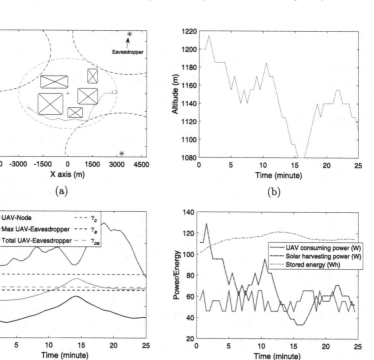

FIGURE 6.4 Simulation results in Environment 2 by the benchmark method. (a) The UAV movement on the xy-plane. A video is available at https://youtu.be/FzNNrJzABzQ. (b) The UAV altitude. (c) The communication and eavesdropping performance. (d) Energy consumption and solar energy harvesting.

boundary of one of the eavesdroppers. Although this does not lead to successful individual eavesdropping by the top eavesdroppers, the two eavesdroppers can recover the captured data by analyzing the combined data. This is shown in Fig. 6.3c. Between 5 and 13 minutes, the eavesdropping performance is above the threshold γ_{ce}. A similar finding can be observed in Fig. 6.4c. From this comparison, we can see that the proposed method can construct a safe trajectory, which ensures communication with the ground node, avoids eavesdropping, and avoids collision with no-fly zones.

6.5 Summary

In this chapter, we considered the problem of securing wireless communication between a solar-powered UAV and a stationary ground node in the presence of stationary eavesdroppers which can carry out the eavesdropping task independently and collaboratively. We developed a 3D UAV trajectory optimization model which minimizes the UAV flight energy expenditure, subject to that the communication between the UAV and the ground node is reliable and secure, and the UAV does deplete the battery during the flight. An RRT trajectory planning scheme was developed. Computer simulations showed that the proposed method guarantees a valid wireless communication link with the ground node, prevents eavesdropping, and avoids no-fly zones.

References

[1] Y. Ota, T. Masuda, K. Araki, M. Yamaguchi, A mobile multipyranometer array for the assessment of solar irradiance incident on a photovoltaic-powered vehicle, Solar Energy 184 (2019) 84–90.
[2] J. Zhang, M. Lou, L. Xiang, L. Hu, Power cognition: enabling intelligent energy harvesting and resource allocation for solar-powered UAVs, in: Future Generation Computer Systems, 2019.
[3] S. Jung, Y. Jo, Y.-J. Kim, Aerial surveillance with low-altitude long-endurance tethered multirotor UAVs using photovoltaic power management system, Energies 12 (7) (2019) 1323.
[4] A.V. Savkin, H. Huang, A method for optimized deployment of a network of surveillance aerial drones, IEEE Systems Journal 13 (4) (2019) 4474–4477.
[5] W. Ejaz, M.A. Azam, S. Saadat, F. Iqbal, A. Hanan, Unmanned aerial vehicles enabled IoT platform for disaster management, Energies 12 (14) (2019) 2706.
[6] D.H. Lee, J.H. Park, Developing inspection methodology of solar energy plants by thermal infrared sensor on board unmanned aerial vehicles, Energies 12 (15) (2019) 2928.
[7] J. Wang, G. Wang, X. Hu, H. Luo, H. Xu, Cooperative transmission tower inspection with a vehicle and a UAV in urban areas, Energies 13 (2) (2020) 326.
[8] A. Mohiuddin, T. Taha, Y. Zweiri, D. Gan, UAV payload transportation via RTDP based optimized velocity profiles, Energies 12 (16) (2019) 3049.
[9] H. Huang, A.V. Savkin, A method for optimized deployment of unmanned aerial vehicles for maximum coverage and minimum interference in cellular networks, IEEE Transactions on Industrial Informatics 15 (5) (2019) 2638–2647.
[10] A.V. Savkin, H. Huang, Range-based reactive deployment of autonomous drones for optimal coverage in disaster areas, IEEE Transactions on Systems, Man, and Cybernetics: Systems (2019) 1–5.
[11] H. Huang, A.V. Savkin, An algorithm of efficient proactive placement of autonomous drones for maximum coverage in cellular networks, IEEE Wireless Communications Letters 7 (6) (2018) 994–997.
[12] J. Lyu, Y. Zeng, R. Zhang, T.J. Lim, Placement optimization of UAV-mounted mobile base stations, IEEE Communications Letters 21 (March 2017) 604–607.
[13] Y. Chen, W. Feng, G. Zheng, Optimum placement of UAV as relays, IEEE Communications Letters 22 (Feb 2018) 248–251.
[14] D. Wang, B. Bai, G. Zhang, Z. Han, Optimal placement of low-altitude aerial base station for securing communications, IEEE Wireless Communications Letters 8 (June 2019) 869–872.
[15] Q. Wang, Z. Chen, W. Mei, J. Fang, Improving physical layer security using UAV-enabled mobile relaying, IEEE Wireless Communications Letters 6 (June 2017) 310–313.
[16] A. Li, Q. Wu, R. Zhang, UAV-enabled cooperative jamming for improving secrecy of ground wiretap channel, IEEE Wireless Communications Letters 8 (1) (2019) 181–184.
[17] Y. Cai, F. Cui, Q. Shi, M. Zhao, G.Y. Li, Dual-UAV-enabled secure communications: joint trajectory design and user scheduling, IEEE Journal on Selected Areas in Communications 36 (Sep. 2018) 1972–1985.
[18] Y. Li, R. Zhang, J. Zhang, L. Yang, Cooperative jamming via spectrum sharing for secure UAV communications, IEEE Wireless Communications Letters (2019).
[19] M. Cui, G. Zhang, Q. Wu, D.W.K. Ng, Robust trajectory and transmit power design for secure UAV communications, IEEE Transactions on Vehicular Technology 67 (Sep. 2018) 9042–9046.

[20] G. Zhang, Q. Wu, M. Cui, R. Zhang, Securing UAV communications via joint trajectory and power control, IEEE Transactions on Wireless Communications 18 (Feb 2019) 1376–1389.

[21] Y. Sun, D. Xu, D.W.K. Ng, L. Dai, R. Schober, Optimal 3D-trajectory design and resource allocation for solar-powered UAV communication systems, IEEE Transactions on Communications 67 (June 2019) 4281–4298.

[22] X. Yuan, Z. Feng, W. Ni, R.P. Liu, J.A. Zhang, W. Xu, Secrecy performance of terrestrial radio links under collaborative aerial eavesdropping, IEEE Transactions on Information Forensics and Security 15 (2020) 604–619.

[23] A.V. Savkin, H. Huang, Optimal aircraft planar navigation in static threat environments, IEEE Transactions on Aerospace and Electronic Systems 53 (Oct 2017) 2413–2426.

[24] T. Inanc, M.K. Muezzinoglu, K. Misovec, R.M. Murray, Framework for low-observable trajectory generation in presence of multiple radars, Journal of Guidance, Control, and Dynamics 31 (6) (2008) 1740–1749.

[25] M. Zabarankin, S. Uryasev, R. Murphey, Aircraft routing under the risk of detection, Naval Research Logistics (NRL) 53 (8) (2006) 728–747.

[26] L.E. Kavraki, P. Svestka, J.-C. Latombe, M.H. Overmars, Probabilistic roadmaps for path planning in high-dimensional configuration spaces, IEEE Transactions on Robotics and Automation 12 (4) (1996) 566–580.

[27] E.W. Dijkstra, et al., A note on two problems in connexion with graphs, Numerische Mathematik 1 (1) (1959) 269–271.

[28] H. Huang, A.V. Savkin, W. Ni, Energy-efficient 3D navigation of a solar-powered UAV for secure communication in the presence of eavesdroppers and no-fly zones, Energies 13 (6) (2020) 1445.

[29] Y. Kang, J.K. Hedrick, Linear tracking for a fixed-wing UAV using nonlinear model predictive control, IEEE Transactions on Control Systems Technology 17 (5) (2009) 1202–1210.

[30] J.A. Duffie, W.A. Beckman, Solar Engineering of Thermal Processes, John Wiley & Sons, 2013.

[31] H.T. Friis, A note on a simple transmission formula, Proceedings of the IRE 34 (5) (1946) 254–256.

Chapter 7

Multiobjective path planning of a solar-powered UAV for secure communication in urban environments with eavesdropping avoidance[☆]

7.1 Motivation

There are different approaches to protecting the wireless communication of UAVs. A common approach is encryption, which encodes a message in such a way that only authorized parties can access it and those who are not authorized cannot. However, if the eavesdropper somehow knows the encoding/decoding schemes used by the transmitter/receiver, it can extract information from the collected data.

In recent years, physical layer security has attracted much attention. The basic idea is to enlarge the received rate at the desired node and reduce that at the eavesdropper so that even if the eavesdropper could receive the transmitted data, it cannot be decoded successfully. Following this approach, the paper [1] considers the optimal deployment of a UAV to secure the communication between the UAV and a ground node in the presence of eavesdroppers. The paper [2] jointly optimizes the trajectory of the UAV and the transmit power of the ground node to avoid eavesdropping. The paper [3] takes into account the NFZ, such as the controlled areas with risks of being targeted by ground-to-air missiles, in the process of path planning. The difference lies in the consideration of the negative impacts of the NFZ. The NFZ in that reference refers to some restricted areas such as some controlled air space, and it does not block the LoS between the UAV and the intended node and that between the UAV and the sun. However, the NFZ considered in the current chapter refers to some tall buildings in urban environments. They may not only block the signal propagation but also prevent the UAV from harvesting energy. Then, the avoidance of the NFZ is similar to conventional obstacle avoidance. Beyond the case with the stationary

---

☆ The main results of the chapter were originally published in Hailong Huang, Andrey V. Savkin, Autonomous navigation of a solar-powered UAV for secure communication in urban environments with eavesdropping avoidance, Future Internet 12 (10) (2020) 170.

Copyright © 2022 Elsevier Inc. All rights reserved.

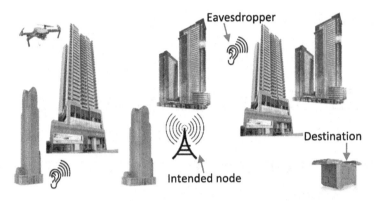

FIGURE 7.1 Illustration of the considered scenario.

ground node and the stationary eavesdroppers, the more challenging situation with a moving ground node and moving eavesdroppers is investigated in [4], and the 3D UAV trajectory planning problem has been studied.

One limitation of many publications on this topic is that they consider a free environment, where the UAV, the ground node, and the eavesdroppers can have LoS [5]. They may not work in urban environments, since the LoS can be blocked by tall buildings [6]. For example, the NFZ considered in [3] refers to some restricted areas such as some controlled air spaces, and they do not block the LoS between the UAV and the intended node and that between the UAV and the sun. Then, the avoidance of the NFZ is similar to conventional collision avoidance. Moreover, the publication [3], as well as many other papers in this area, do not consider scenarios with some tall buildings in urban environments which may not only block the signal propagation but also prevent the UAV from harvesting energy. Another limitation is that energy efficiency has not been comprehensively considered in these publications. Solar-powered UAVs have been employed to prolong the operation duration. However, the negative impacts of the urban environment on energy harvesting has not been well considered.

This chapter considers using a solar-powered UAV to secure the wireless communication with an intended node in the presence of eavesdroppers in an urban environment (Fig. 7.1). The term of securing the wireless communication means that the intended ground node is able to decode the data sent by the UAV while the eavesdroppers cannot. The aforementioned limitations are taken into account in this chapter. Specifically, we consider a given urban environment, where the positions and shapes of the buildings are known. Any building is modeled by a prism. Then, the UAV should fly outside the prisms. Additionally, given the positions of the UAV and the ground node, we can easily verify whether they have LoS. Similarly, the LoS between the UAV and the sun can be checked, which makes the estimation of the harvested energy amount accurate.

We focus on the UAV path planning problem by taking into account solar energy harvesting and instantaneous communication protection. We formulate a multiobjective optimization model to maximize the residual energy of the UAV at the end of the mission, maximize the time period in which the communication between the UAV and the intended node is secure, and minimize the time to arrive at the destination. To address this problem, we propose an RRT-based [7] scheme. This RRT-based scheme captures the nonlinear UAV motion model, and is also computationally efficient considering the randomness nature. From the generated tree, a set of possible paths that end up in a given region, such as the UAV depot (this is to facilitate the UAV collection after the mission; see Fig. 7.1), can be found. For each possible path, we can verify the security of the wireless communication and compute the overall energy consumption, the harvested amount, and the time to complete the path. We then pick the path with the maximum value of a joint metric from the feasible paths. The main contributions of this chapter are the new problem formulation and the new RRT-based path planning method. We test the effectiveness of the proposed method via computer simulations. By comparing with a benchmark method, the proposed method guarantees secure wireless communication with the intended node and prevents eavesdropping in a large time window.

7.1.1 Related work

The problem considered in this chapter belongs to classical path planning and secure communication. In this subsection, we first briefly review the typical approaches and then clarify the differences between our work and existing publications.

Graph search-based approaches are the well-studied ones for global path planning. They generally need the environment to be modeled by a graph, where a possible link between two vertices of the graph is associated with a weight indicating the cost or time to traverse the link. Then, many popular graph search algorithms, such as Dijkstra's algorithm [8] and the A* algorithm [9], can be applied to compute the shortest path between a start vertex and a destination vertex.

Tangent graph-based approaches are another type of tools to compute globally optimal paths. For a nonholonomic robot, say a Dubins car [10], its angular speed is upper-bounded (its minimum turning radius is lower-bounded). Then, given the obstacles in an environment, the obstacles can be modeled by circles whose radius should be no smaller than the robot minimum turning radius. Furthermore, a set of tangent lines can be added to link a pair of obstacle circles. The generated graph is called the tangent graph and the robot's path in this tangent graph consists of a number of straight-line segments and a number of arcs on the obstacle circle [11,12].

Optimization algorithms are often used for the path planning problem. Typical optimization algorithms that have been used include but are not limited to PSO [13], the genetic algorithm [14], differential evolution [15], and the gravitational search algorithm [16]. The methods are in general based on a pool of possible trajectories and follow some updating rules until the optimal path is obtained. A shortcoming of this class of approaches is that they may converge slowly in normal path planning problems.

The RRT scheme, a randomized method, usually performs well in terms of computational efficiency [7]. This method is based on the construction of a random tree of possible actions connecting the start position and the destination. When a node of the tree reaches the destination, a feasible path from the start point to the destination can then be identified. Thanks to its computational efficiency, the RRT scheme is promising to be applied in online path planning. The main task is to dynamically grow and manage the tree. For example, when the destination changes or when mobile obstacles come into proximity, some new nodes may be added to the tree and some existing nodes of the tree may be cut to avoid collision [17,18]. Moreover, beyond the general RRT framework, some heuristic methods have been proposed to guide the tree growth [19,20], which may generate nearly optimal paths while remaining with a probabilistic planning setting.

The considered application scenario is about secure wireless communication. Secure communication is not a new issue in wireless communication, and there are approaches to protecting security. However, the wireless communication of UAVs brings new challenges. The most important feature is the high probability of LoS with not only the intended node but also malicious nodes (i.e., eavesdroppers). As mentioned, some strategies have been proposed to increase the security capacity of UAV wireless communication. The papers [2–4] focus on UAV mobility management to increase the received rate at the intended node and decrease that at the eavesdroppers. Besides this approach, another class of approaches employs UAVs as jammers [21–23]. Specifically, the paper [21] considers the optimization of the UAV jammer's trajectory and transmission power. The paper [22] further takes into account the user scheduling to maximize the secrecy rate. The paper [23] proposes a strategy where two UAVs send confidential messages to their respective intended node by sharing the same spectrum. A cooperation strategy is designed to maximize the system secrecy rate.

The path planning problem considered in this paper is for a UAV whose path is 3D. However, most graph search-based approaches and tangent graph-based approaches focus on 2D path planning problems. Additionally, the considered application scenario, i.e., securing the wireless communication between the solar-powered UAV and a ground node in the presence of eavesdroppers in urban environments, introduces some new requirements on the UAV path. Specifically, the UAV should ensure that the ground node can successfully decode the sent data while the eavesdroppers cannot decode the sent data. The main results of the chapter were originally published in [24].

The rest of this chapter is organized as follows. Section 7.2 presents the system models and states the problem of interest. Section 7.3 presents the proposed path planning method. Computer simulations are shown in Section 7.4. Finally, a summary is given in Section 7.5.

TABLE 7.1 Symbols and meanings.

Symbol	Meaning
$p(t)$	Position of the UAV
$v(t)$	Linear speed on the xy-plane
$w(t)$	Angular speed on the xy-plane
$u(t)$	Vertical speed
$\theta(t)$	Heading of the UAV on the xy-plane
$D(t)$	Drag of the UAV
$f(t)$	Thrust of the UAV
m	The mass of the UAV
P^{sun}	Energy harvesting power of the UAV
$P(t)$	Energy consuming power of the UAV
$Q(t)$	Residual energy of the UAV
$\mathcal{T}$	The random tree
$\mathcal{R}$	The set of paths in $\mathcal{T}$

7.2   Problem statement

In this section, we first present the system model and then formulate the problem of interest. The main symbols used in the paper and their meanings are summarized in Table 7.1.

7.2.1   UAV model

We consider a solar-powered UAV that flies in an urban environment. We denote $p(t) = [x(t), y(t), z(t)]$ as the coordinates of the UAV at time t. The following model is used to describe the UAV's motion[1]:

$$
\begin{cases}
\dot{x}(t) = v(t)\cos(\theta(t)), \\
\dot{y}(t) = v(t)\sin(\theta(t)), \\
\dot{\theta}(t) = w(t), \\
\dot{z}(t) = u(t), \\
\dot{u}(t) = \frac{F(t)-D(t)}{m},
\end{cases}
\tag{7.1}
$$

where $\theta(t)$ is the heading of the UAV with respect to the x-axis, $v(t)$, $\omega(t)$, and $u(t)$ are its linear horizontal, angular, and vertical speeds, respectively, m is the mass of the UAV, and $D(t)$ and $F(t)$ are the drag and thrust forces, respectively. The model (7.1) and its slight modifications have been widely used to describe the motion of aircraft, wheeled robots, and missiles (see, e.g., [25,26]).

---

[1] This 3D UAV motion model is extended from a 2D model [25] where a forward speed and an angular speed are the control inputs. Beyond these two control inputs, we consider the vertical speed such that the UAV can adjust its altitude.

The drag force is computed as follows [27]:

$$D(t) = \frac{1}{2}\rho C_D A v^2(t), \tag{7.2}$$

where ρ is the air density, A is the area of the solar cells of the UAV, $C_D = C_{D0} + \frac{C_L^2}{\varepsilon \pi R_a}$ is the coefficient of drag, C_{D0} is the parasitic drag coefficient, $C_L = \frac{2mg}{\rho A v^2(t)}$ is the coefficient of lift, R_a is the aspect ratio of the wing, and ε is the Oswald efficiency factor.

Moreover, the following constraints are enforced on the UAV's movement at any time t:

$$\begin{cases} Z^{min} \leq z(t) \leq Z^{max}, \\ 0 < v(t) \leq V^{max}, \\ -\Omega^{max} \leq w(t) \leq \Omega^{max}, \\ -U^{max} \leq u(t) \leq U^{max}, \\ 0 \leq F(t) \leq F^{max}, \end{cases} \tag{7.3}$$

where the constants $0 < Z^{min} < Z^{max}$ specify the allowed deployment altitude, the constants V^{max}, Ω^{max}, and U^{max} specify the bounds of the corresponding speeds of the UAV, and the constant F^{max} is the maximum allowed thrust force.

7.2.2 Energy harvesting and consuming

The solar-powered UAV can harvest energy from the sun. Let P^{sun} denote the harvesting, which can be computed by [27]

$$P^{sun} = \eta A \cos\phi, \tag{7.4}$$

where η is the efficiency of the solar cell and ϕ is the incidence angle. Here ϕ depends on the azimuth angle α_z and the elevation angle α_e of the sun, and α_z and α_e are time-varying in the daytime. Therefore, ϕ also varies with time.

When flying, the UAV also consumes energy. Let $P(t)$ denote the energy consumption power, calculated by

$$P(t) = \frac{F(t)v(t)}{\eta_{prop}}, \tag{7.5}$$

where η_{prop} is the efficiency of the propeller and $v(t) = \sqrt{v^2(t) + u^2(t)}$.

Let $Q(t)$ denote the residual energy of the battery of the UAV, which is upper-bounded by the capacity Q^{max}. We have

$$\dot{Q}(t) = P^{sun} - P(t). \tag{7.6}$$

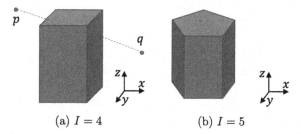

(a) $I = 4$ (b) $I = 5$

FIGURE 7.2 Prisms.

7.2.3 No-fly zone

There are some buildings in the considered environment, and they occupy some space where the UAV is prohibited. We call this space the NFZ. We model each building by the smallest prism enclosing this building. Each prism has two parallel and congruent bases and a number of flat sides (Fig. 7.2). The shape of the bases depends on the building. All the prisms are assumed to have one base on the xy-plane and the other above the xy-plane. All the sides of the prisms are assumed to be perpendicular to the xy-plane. Each prism can be characterized by three parameters: (1) an I-by-2 matrix Ξ, (2) an I-by-1 vector ξ, and (3) a scalar h. The integer I depends on the shape of the base. Any point (x, y, z) inside a prism satisfies the following condition:

$$\begin{cases} \Xi \begin{bmatrix} x \\ y \end{bmatrix} \leq \xi, \\ 0 \leq z \leq h. \end{cases} \tag{7.7}$$

Given the environment, all the buildings have the corresponding Ξ, ξ, and h. At any time, the UAV must not be at a position that satisfies (7.7). The problem of avoidance of NFZ is similar in spirit to the problem of avoiding collisions with steady obstacles (see, e.g., [28,29] and the references therein).

7.2.4 Line-of-sight

In addition to being NFZs, tall buildings may block the wireless channels. Moreover, they may create some shadow regions where the solar-powered UAV cannot harvest energy from the sun. Thus, it is necessary to have an evaluation model to verify whether the LoS condition is satisfied between the UAV and the ground node and between the UAV and the sun.

Suppose that the two entities are located at p and q, respectively. The straight-line segment between p and q can be written as follows:

$$\begin{cases} x = x_q + \alpha\tau, \\ y = y_q + \beta\tau, \\ z = z_q + \gamma\tau, \\ \min\{x_p, x_q\} \le x_q + \alpha\tau \le \max\{x_p, x_q\}, \end{cases} \tag{7.8}$$

where $q = (x_q, y_q, z_q)$, $p = (x_p, y_p, z_p)$, and $\frac{\vec{pq}}{\|\vec{pq}\|} = (\alpha, \beta, \gamma)$. The last inequality specifies the range of the free variable τ.

Whether the LoS between p and q is blocked by a prism depends on whether the line segment (7.8) and the corresponding prism (7.7) have any intersection points. To this end, we can solve Eqs. (7.7) and (7.8) for each prism. If there are solutions to any prism, the LoS is blocked (Fig. 7.2); otherwise, the LoS between p and q is not blocked.

Note that to verify if the LoS between the UAV and the sun is blocked, we need to have the sun's location. Let V be the unit vector representing the sunlight direction. With the azimuth angle α_z and the elevation angle α_e of the sun, the vector $V = [\cos\alpha_e \cos\alpha_z, \cos\alpha_e \sin\alpha_z, -\sin\alpha_z]^T$ [30]. We can imagine that the sun is located at $q_{sun} = p - V\tau$, where τ takes a large value so that the sun is very far from p. Let $b(t)$ be a binary variable: $b(t) = 1$ if the UAV and the sun have LoS, and $b(t) = 0$ otherwise. With the symbol $b(t)$, the residual energy of the UAV should be calculated according to the following equation:

$$\dot{Q}(t) = P^{sun}b(t) - P(t). \tag{7.9}$$

For the safety of operation, at any time, the residual energy of UAV i cannot be smaller than the threshold Q^{min}:

$$Q(t) \ge Q^{min}. \tag{7.10}$$

7.2.5 Secure communication

Let (x_i, y_i) be the position of the ground node i, and $i = 0, \ldots, N$. When $i = 0$, the node refers to the intended ground node, while when $i = 1, \ldots N$, the node is an eavesdropper. The intended node is a legitimate node, and it makes sense for the UAV to know its location. For the locations of eavesdroppers, following [31], we assume that they can be measured by an optical camera or a synthetic aperture radar. Here, N is the number of eavesdroppers. Let $d_i(t)$ denote the Euclidean distance between the UAV and node i at time t:

$$d_i(t) = \sqrt{(x(t) - x_i)^2 + (y(t) - y_i)^2 + z(t)^2}.$$

Denote $\kappa(d)$ as a function measuring the communication performance. In the free space, $\kappa(d)$ is a decreasing function of d. In the well-known Friis formula [32], when the transmitter and the receiver are d apart from each other with LoS, the received SNR is $\kappa(d) = \frac{a}{d^2}$, where a is a given positive constant. Let the binary variable $l_i(t)$ indicate whether the UAV has LoS with node i at t: $l_i(t) = 1$ if they have LoS, and $l_i(t) = 0$ otherwise. Let $\kappa(d_i(t), l_i(t))$ denote the SNR between the UAV and node i. Now $\kappa(d_i(t), l_i(t))$ is defined as follows:

$$\kappa(d_i(t), l_i(t)) = \frac{al_i(t)}{d_i^2(t)}. \tag{7.11}$$

To secure the wireless communication between the UAV and the intended node, a set of constraints are considered below. The first requirement is that the received SNR at the intended node is no smaller than a threshold γ_c:

$$\kappa(d_0(t), l_0(t)) \geq \gamma_c. \tag{7.12}$$

This is to ensure that the node can successfully recover the received data at time t. Secondly, the received SNR at each eavesdropper must be no larger than a threshold γ_e ($\gamma_e < \gamma_c$):

$$\kappa(d_i(t), l_i(t)) \leq \gamma_e, \tag{7.13}$$

for all $i = 1, \ldots, N$. This is to ensure that any eavesdropper i cannot recover the captured data.

In general, the constraint (7.12) requires the UAV to be close enough to the intended node and have LoS with the node, whereas the constraint (7.13) requires the UAV to be far enough from the eavesdroppers or have no LoS with the eavesdropper. We note that in cases where several eavesdroppers are located closely, they may block the path for the UAV, and there may not exist a feasible path satisfying (7.12) and (7.13). To avoid being eavesdropped, the UAV stops sending data in these cases. We also note that the buildings can reflect wireless signals. This can make the received SNR larger than the value computed by (7.11). Since the non-LoS propagation model is complex and out of the scope of this paper, we do not go far in this direction. It is worth pointing out that if such a model is available, our method can be extended straightforwardly by updating Eq. (7.11).

7.2.6 Problem statement

Suppose the UAV operation period is $[0, T]$. The UAV's initial position is $p(0)$ and the initial energy is $Q(0)$. Let $\mathcal{D}$ denote a given bounded and connected destination set. Let T^* denote the time instant at which the UAV arrives at $\mathcal{D}$. In other words, T^* is the time needed to complete the flight. During the period $[0, T^*]$, the UAV transmits confidential data to the intended ground node if eavesdropping can be avoided. The path planning problem under investigation is to jointly optimize three objectives: minimizing T^*, maximizing the final residual energy $Q(T^*)$, and maximizing the total time the UAV can securely transmit data by finding a sequence of control inputs $v(t), \omega(t), u(t)$. Formally, the problem is formulated as follows:

$$\min_{v(t),\omega(t),u(t)} T^*, \tag{7.14}$$

$$\max_{v(t),\omega(t),u(t)} Q(T^*), \tag{7.15}$$

$$\max_{v(t),\omega(t),u(t)} \int_0^{T^*} l_0(t)dt \tag{7.16}$$

subject to

$$Q^{min} \le Q(t) \le Q^{max}, \ \forall t \in [0, T], \tag{7.17}$$

$$(x(T), y(T)) \in \mathcal{D}, \tag{7.18}$$

(7.3), (7.12) and (7.13) hold, while (7.7) does not hold for any prism at any time.

Different from the classic path planning problem which targets a collision-free path for a mobile robot to safely and quickly arrive at the destination [33], the problem of interest also aims at maximizing the residual energy of the solar-powered UAV and total time duration in which the UAV can securely communicate with the intended node. In the considered scenario, taking the aforementioned factors into account together is necessary. Suppose we do not consider the flight time. Then, the UAV trajectory may be very long. If we do not consider the harvested energy amount, the UAV trajectory may have a large part in the shadow created by the buildings.

7.3 RRT-based path planning

The problem under consideration cannot be solved by convex optimization tools. The reason lies in the nonconvex constraints of (7.12) and (7.13), the nonsmoothness of the residual energy model (7.9) due to the shadow regions created by the buildings, and the nonlinear UAV model (7.1). Moreover, the minimization of the flight time and the optimization of the UAV path are difficult to decouple. To overcome these difficulties, we develop an RRT-based path planning method. As mentioned, the RRT method can easily tackle the nonlinear UAV model, secure communication requirements, and the NFZ by randomly generating samples.

Let $\mathcal{T}$ be the random tree, which consists of a number of vertices. Each vertex is associated with the status of the UAV, including its position, heading, linear speed, vertical speed, angular speed, and whether it has LoS with the sun, the intended node, and the eavesdroppers. It is assumed that the UAV knows the locations of the intended node and the eavesdroppers and the sun's direction vector; otherwise, it is impossible to guarantee secure communication and verify the LoS with these entities. The operation period is discretized by a sampling interval δ. This sampling interval restricts the movement of the UAV in a single step.

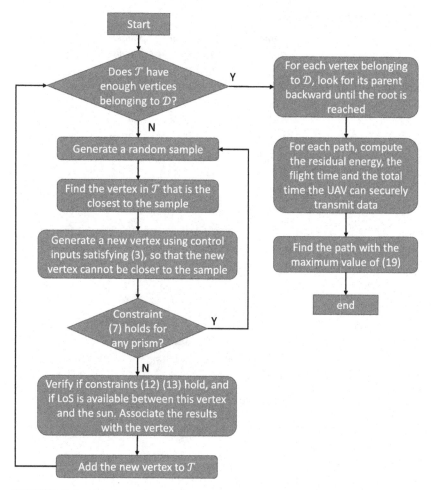

FIGURE 7.3 Flow chart of the RRT-based path planning method.

Following the framework of RRT, we can keep generating random samples between the allowed altitudes Z^{min} and Z^{max} in the environment. For any sample, we look for the closest vertex in the tree $\mathcal{T}$ to this sample. We generate a new vertex with appropriate control inputs satisfying (7.3), so that the new vertex cannot be closer to the sample after a movement in the interval δ. The new vertex is added to the tree $\mathcal{T}$ if constraint (7.7) does not hold for any prism in the considered environment. Otherwise, this vertex is deleted. If the candidate vertex is added to the tree $\mathcal{T}$, we further verify whether constraints (7.12) and (7.13) hold and if the LoS exists between this vertex and the sun. When a given number of vertices fall into the destination set $\mathcal{D}$, we can terminate the tree growing process. From the generated random tree $\mathcal{T}$, we can find a num-

TABLE 7.2 Parameters.

Parameter	Value	Parameter	Value
V_{max}	20 m/s	Ω_{max}	1 rad/s
U_{max}	2 m/s	Z_{max}	100 m
Z_{min}	20 m	V	[0.8, 0, −0.6]
E_{max}	200	E_{min}	20
δ	1 s	T	150 s
m	0.5 kg	ρ	1.29 kg/m^3
A	0.5 m^2	C_{D_0}	0.011
ϵ	0.1	R_a	10
η_{prop}	0.1	g	9.8 kg^{-2} m^2
a	2.5×10^5	$p(0)$	[0, 0, 60]
γ_c	1	γ_e	0.8

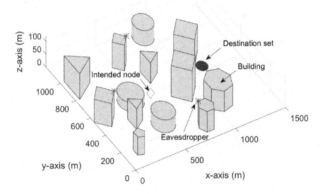

FIGURE 7.4 The simulated urban environment.

ber of paths to evaluate. For finding the paths, we consider each of the vertices falling into the destination set $\mathcal{D}$. Specifically, from this vertex, we look backwards to identify its parent vertex until the root (i.e., the vertex indicating the initial position of the UAV) is reached. Let $\mathcal{R}$ denote the set of paths found by the above procedure. These paths are feasible for the UAV, since any vertex is generated by appropriate control inputs. Additionally, any vertex is outside the prisms (which is safe for the UAV). The last step is to make the final selection of the path. As the considered problem is a multiobjective problem, we select the path that maximizes the following metric:

$$\max_{\mathcal{P} \in \mathcal{R}} \frac{Q(\mathcal{P}) \int_0^{T(\mathcal{P})} l_0(t)dt}{T(\mathcal{P})}, \qquad (7.19)$$

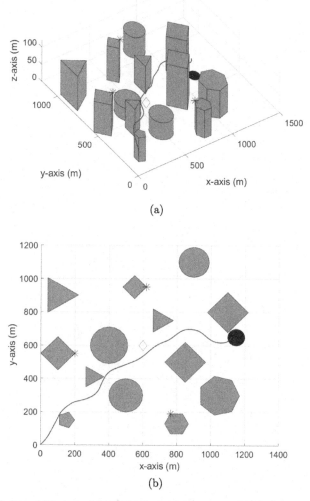

(a)

(b)

FIGURE 7.5 3D and 2D views of the UAV trajectory by the proposed method.

where $T(\mathcal{P})$ represents the UAV flight time following the path $\mathcal{P}$, $Q(\mathcal{P})$ represents the residual energy of the UAV, and $\int_0^{T(\mathcal{P})} l_0(t)dt$ gives the total time in which the transmission by the UAV is secure. It is worth pointing out that (7.19) is not the only metric to evaluate a path. An alternative can be a linear combination of the three objectives. Since we need to choose two weights in that kind of metric, we use (7.19) in this paper to evaluate a path. After evaluating all the possible paths, we can finally find the one having the maximum value of (7.19). All the procedures of this method are summarized in Fig. 7.3.

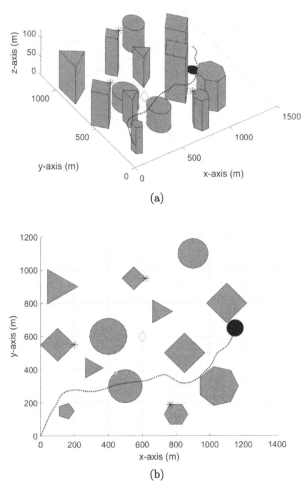

FIGURE 7.6 3D and 2D views of the UAV trajectory by the compared method.

It is worth pointing out that although the currently considered scenario involves a stationary intended node, the proposed method can also be used for the cases with a moving intended node. Thanks to the computational efficiency, the task of adding new vertices and removing old vertices can be done online.

7.4 Simulation results

The effectiveness of the proposed method is shown in this section via computer simulations in MATLAB®. The parameters used in the simulations are shown

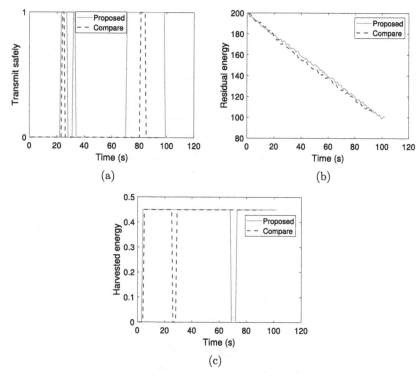

FIGURE 7.7 Simulation results of when the UAV can securely transmit data, the residual energy, and the harvested energy.

in Table 7.2. We create an urban environment with a number of buildings, as shown in Fig. 7.4. There is an intended node marked by a diamond. There are three eavesdroppers deployed on the roof of some buildings, marked by the black stars. The destination set $\mathcal{D}$ is marked by a blue disk. Our UAV starts from $p(0)$. Applying the proposed scheme, the UAV path is generated in less than one second. We show the UAV path in Fig. 7.5. The results of the residual energy, the flight time, and the time window in which the communication is secure are plotted in Fig. 7.7. The UAV completes the flight to reach the destination set $\mathcal{D}$ in 102 seconds, and the residual energy is 100 units. In 34% of this period, the UAV can securely communicate with the intended node.

For comparison, we replace the path selection metric with the minimization of the flight time and apply it to the above case. The obtained UAV path by this compared method is shown in Fig. 7.6. The corresponding results are shown in Fig. 7.7. Following this path, the UAV arrives at the destination set in 92 seconds (Fig. 7.7b). The residual energy is 110 units. This path outperforms the above one in these two terms. However, as seen from Fig. 7.7a, this path enables

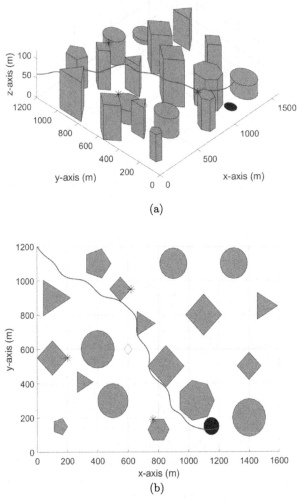

FIGURE 7.8 3D and 2D views of the UAV trajectory by the proposed method in another urban environment.

secure communication in only 6% of the period, which is much less than for the proposed method. Fig. 7.7c shows the harvested energy during the flight.

The proposed method is not restricted to any environment. We created another urban environment and then applied the proposed method and the compared method to construct the UAV path. The UAV path obtained by the proposed method is shown in Fig. 7.8 with both 2D and 3D views. The time window in which the UAV can securely transmit and the residual energy are shown in

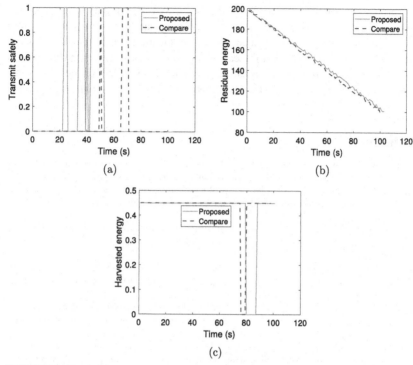

FIGURE 7.9 Simulation results of when the UAV can securely transmit data, the residual energy, and the harvested energy.

Fig. 7.9. Similar to the results in Fig. 7.7, the proposed method obtains the UAV path that has a similar flight time and similar residual energy, but a much larger time window for secure communication than the compared method.

7.5 Summary

In this chapter, we considered the application of using a solar-powered UAV for securing communication with a ground node in the presence of eavesdroppers in urban environments. A path planning problem was formulated with the objective of jointly maximizing the residual energy of the solar-powered UAV at the end of the mission, maximizing the time period in which the UAV can securely communicate with the intended node, and minimizing the time to reach the destination. Special attention was paid to the impact of the buildings in the urban environments, which may block the transmitted signals and also create some shadow region where the UAV cannot harvest energy. An RRT-based path planning scheme was presented. The effectiveness of the proposed scheme was tested via computer simulations. The comparison with a general selection of the path to minimize the energy consumption showed that the path generated by the proposed scheme enables the UAV to securely transmit data to the intended node in a larger time window. One of our future research directions is to explore a UAV team for the purpose of securing communication, where one of the UAVs transmits data and others jam the eavesdroppers. Another research direction is to extend the current work to dynamic environments with either the mobile intended node or moving obstacles in the environment. Then, we will focus on the replanning of the UAV path.

References

[1] D. Wang, B. Bai, G. Zhang, Z. Han, Optimal placement of low-altitude aerial base station for securing communications, IEEE Wireless Communications Letters 8 (June 2019) 869–872.

[2] M. Cui, G. Zhang, Q. Wu, D.W.K. Ng, Robust trajectory and transmit power design for secure UAV communications, IEEE Transactions on Vehicular Technology 67 (Sep. 2018) 9042–9046.

[3] H. Huang, A.V. Savkin, W. Ni, Energy-efficient 3D navigation of a solar-powered UAV for secure communication in the presence of eavesdroppers and no-fly zones, Energies 13 (6) (2020) 1445.

[4] A.V. Savkin, H. Huang, W. Ni, Securing UAV communication in the presence of stationary or mobile eavesdroppers via online 3D trajectory planning, IEEE Wireless Communications Letters 9 (8) (2020) 1211–1215.

[5] X. Yuan, Z. Feng, W. Ni, R.P. Liu, J.A. Zhang, W. Xu, Secrecy performance of terrestrial radio links under collaborative aerial eavesdropping, IEEE Transactions on Information Forensics and Security 15 (2020) 604–619.

[6] H. Huang, A.V. Savkin, M. Ding, M.A. Kaafar, Optimized deployment of drone base station to improve user experience in cellular networks, Journal of Network and Computer Applications 144 (2019) 49–58.

[7] S.M. LaValle, J.J. Kuffner Jr., Randomized kinodynamic planning, The International Journal of Robotics Research 20 (5) (2001) 378–400.

[8] E.W. Dijkstra, et al., A note on two problems in connexion with graphs, Numerische Mathematik 1 (1) (1959) 269–271.

[9] P.E. Hart, N.J. Nilsson, B. Raphael, A formal basis for the heuristic determination of minimum cost paths, IEEE Transactions on Systems Science and Cybernetics 4 (2) (1968) 100–107.

[10] L.E. Dubins, On curves of minimal length with a constraint on average curvature, and with prescribed initial and terminal positions and tangents, American Journal of Mathematics 79 (3) (1957) 497–516.

[11] A.V. Savkin, M. Hoy, Reactive and the shortest path navigation of a wheeled mobile robot in cluttered environments, Robotica 31 (2) (2013) 323.

[12] A.V. Savkin, H. Huang, Optimal aircraft planar navigation in static threat environments, IEEE Transactions on Aerospace and Electronic Systems 53 (Oct 2017) 2413–2426.

[13] Y. Zhang, D.-W. Gong, J.-H. Zhang, Robot path planning in uncertain environment using multi-objective particle swarm optimization, Neurocomputing 103 (2013) 172–185.

[14] V. Roberge, M. Tarbouchi, G. Labonté, Comparison of parallel genetic algorithm and particle swarm optimization for real-time UAV path planning, IEEE Transactions on Industrial Informatics 9 (1) (2012) 132–141.

[15] Y. Shen, Y. Wang, Operating point optimization of auxiliary power unit using adaptive multi-objective differential evolution algorithm, IEEE Transactions on Industrial Electronics 64 (1) (2016) 115–124.

[16] E. Rashedi, H. Nezamabadi-Pour, S. Saryazdi, GSA: a gravitational search algorithm, Information Sciences 179 (13) (2009) 2232–2248.

[17] D. Ferguson, A. Stentz, Anytime RRTs, in: 2006 IEEE/RSJ International Conference on Intelligent Robots and Systems, IEEE, 2006, pp. 5369–5375.

[18] M. Otte, E. Frazzoli, RRTX: asymptotically optimal single-query sampling-based motion planning with quick replanning, The International Journal of Robotics Research 35 (7) (2016) 797–822.

[19] I. Ko, B. Kim, F.C. Park, VF-RRT: introducing optimization into randomized motion planning, in: 2013 9th Asian Control Conference (ASCC), 2013, pp. 1–5.

[20] J. Wang, M.Q. Meng, O. Khatib, EB-RRT: optimal motion planning for mobile robots, IEEE Transactions on Automation Science and Engineering (2020) 1–11.

[21] A. Li, Q. Wu, R. Zhang, UAV-enabled cooperative jamming for improving secrecy of ground wiretap channel, IEEE Wireless Communications Letters 8 (1) (2019) 181–184.

[22] Y. Cai, F. Cui, Q. Shi, M. Zhao, G.Y. Li, Dual-UAV-enabled secure communications: joint trajectory design and user scheduling, IEEE Journal on Selected Areas in Communications 36 (9) (2018) 1972–1985.

[23] Y. Li, R. Zhang, J. Zhang, L. Yang, Cooperative jamming via spectrum sharing for secure UAV communications, IEEE Wireless Communications Letters 9 (3) (2020) 326–330.

[24] H. Huang, A.V. Savkin, Autonomous navigation of a solar-powered UAV for secure communication in urban environments with eavesdropping avoidance, Future Internet 12 (10) (2020) 170.

[25] Y. Kang, J.K. Hedrick, Linear tracking for a fixed-wing UAV using nonlinear model predictive control, IEEE Transactions on Control Systems Technology 17 (5) (2009) 1202–1210.

[26] H. Li, A.V. Savkin, Wireless sensor network based navigation of micro flying robots in the industrial internet of things, IEEE Transactions on Industrial Informatics 14 (8) (2018) 3524–3533.

[27] A.T. Klesh, P.T. Kabamba, Solar-powered aircraft: energy-optimal path planning and perpetual endurance, Journal of Guidance, Control, and Dynamics 32 (4) (2009) 1320–1329.

[28] A.S. Matveev, H. Teimoori, A.V. Savkin, A method for guidance and control of an autonomous vehicle in problems of border patrolling and obstacle avoidance, Automatica 47 (3) (2011) 515–524.

[29] M. Hoy, A.S. Matveev, A.V. Savkin, Algorithms for collision-free navigation of mobile robots in complex cluttered environments: a survey, Robotica 33 (3) (2015) 463–497.

[30] J. Wu, H. Wang, N. Li, P. Yao, Y. Huang, Z. Su, Y. Yu, Distributed trajectory optimization for multiple solar-powered UAVs target tracking in urban environment by adaptive grasshopper optimization algorithm, Aerospace Science and Technology 70 (2017) 497–510.

[31] G. Zhang, Q. Wu, M. Cui, R. Zhang, Securing UAV communications via joint trajectory and power control, IEEE Transactions on Wireless Communications 18 (2) (2019) 1376–1389.

[32] H.T. Friis, A note on a simple transmission formula, Proceedings of the IRE 34 (5) (1946) 254–256.

[33] R.A. Saeed, D.R. Recupero, Path planning of a mobile robot in grid space using boundary node method, in: Proceedings of the 16th International Conference on Informatics in Control, Automation and Robotics, 2019, pp. 159–166.

Chapter 8

Reactive deployment of UAV base stations for providing wireless communication services[☆]

8.1 Motivation

UAVs are used in various applications including providing wireless communication to ground cellular users or vehicles [1,2], collecting sensory data from WSNs [3], monitoring and surveillance [4,5], and natural disaster management [6,7]. For these applications, the efficient deployment of UAVs is a critical issue, especially when mobility is involved, for example, in scenarios with mobile cellular users in communication support tasks, or with mobile targets in surveillance tasks and with mobile sensors in WSNs.

There are several important issues to be addressed in the problems of deploying UAVs. The first one is the issue of coverage. Regardless of application scenarios, the UAVs should deliver a good coverage of the objects of interest (such as cellular users, ground robots, sensor nodes, etc.). Secondly, the UAVs should form a connected network. For example, in the problems of monitoring and surveillance, the UAVs need to stay connected with ground base stations. More precisely, each UAV has to maintain a valid path to at least one ground base station so that the sensory data can be reported to the base station timely. Another aspect is energy capacity. Since many of the existing commercial UAVs are powered by onboard batteries, their operation time is limited. Therefore, efficient use of battery energy resources is a key issue for networks of UAVs.

There is a large number of publications on coverage by UAVs. One existing approach aims at minimizing the number of UAVs to cover a given set of objects [8–11]. Some other publications studied the problem of maximizing the coverage of objects by a given set of UAVs [4,12]. In general, the first group of papers requires global knowledge of the objects' locations, and therefore, most of them provide offline and centralized algorithms. The second group focuses on dynamic scenarios that do not require a priori knowledge of objects' locations.

☆ The main results of the chapter were originally published in Hailong Huang, Andrey V. Savkin, Reactive 3D deployment of a flying robotic network for surveillance of mobile targets, Computer Networks 161 (2019) 172–182. Permission from Elsevier for reuse was obtained.

Copyright © 2022 Elsevier Inc. All rights reserved.

In such scenarios, the goal is to reactively navigate the UAVs to achieve better coverage.

In this chapter, a UAV network is used to provide cellular service to users in an area of interest for a period of time, and the UAVs play the role of aerial base stations. Specifically, we consider the situation when the number of UAVs is not sufficient to cover the whole area, and we need to deploy them in some optimized way to maximize the coverage. This case is quite practical because in many real-world applications, existing UAVs are not able to cover the whole area and the number and the distribution of users are not known in advance. Also, due to the limited capacity of the onboard battery, a UAV may not operate for the whole period. A practical application of the considered problem is serving users in an area of interest during some specific period of time such as the gathering event on New Year's Eve in the Harbour Bridge area in Sydney. For the problem under consideration, a reactive algorithm to navigate each UAV in 3D space based on instant sensory information and some shared information with nearby UAVs is developed. Unlike many existing publications aiming at minimizing the number of robots or maximizing the number of covered users, the objective here is to simultaneously maximize the network lifetime and the coverage of users. A novel optimization problem taking into account these two aspects together with energy and connectivity constraints is formulated. Furthermore, a decentralized algorithm to solve this problem is developed. Simulations are carried out to evaluate the performance of the proposed algorithm and compare it with some baseline methods. The main results of the chapter were originally published in [13].

The rest of the chapter is organized as follows. Section 8.2 presents the system model and formally states the studied optimization problem. Section 8.3 presents a distributed navigation algorithm. Section 8.4 gives some simulation results to demonstrate the performance of the proposed method. Finally, Section 8.5 concludes the chapter together with some future research directions.

8.2 Problem statement

As mentioned, we focus on the problem of maximizing the user coverage as well as the network lifetime for a given set of UAVs considering the energy and connectivity constraints. We will start with the basic models for connectivity and energy consumption and then we will move to the coverage model. Finally, we formally state the considered problem and analyze its complexity. The main notations used herein are summarized in Table 8.1.

The system consists of a UAV network with n UAVs labeled $i = 1, 2, \ldots, n$ and m ground base stations labeled $l = 1, \ldots, m$. The operation period is $[0, \mathcal{T}]$, which is discretized into T slots. Let $t = 1, \ldots, T$ be the index and the corresponding time slot is $[\frac{\mathcal{T}}{T}(t - 1), \frac{\mathcal{T}}{T}t]$. This system is to serve mobile users with equal importance level in a 2D area of interest, denoted by $\mathcal{D}$. To account for different importance levels of users, a weighting of each user can be added

TABLE 8.1 Notations and descriptions.

Notation	Description
n	The number of UAVs
m	The number of base stations
$\mathcal{D}$	The area of interest
$\mathcal{S}$	The 3D UAV deployment space
U	The set of users
$\mathcal{T}$	The monitoring period
$\mathcal{G}$	The given connectivity graph
B_l	The 2D location of base station l
$u(t)$	The 2D location of user u at time t
$P_i(t)$	The 3D position of UAV i in time slot t
$\mathcal{P}_i(t)$	The set of reachable positions from $P_i(t)$
$E_i(t)$	The residual energy level of robot i
$v_i(t)$	The speed of robot i
$s_i(t)$	The total distance robot i has flown
$r_i(t)$	The coverage radius of robot i
$N_i(t)$	The number of users robot i covers
$U_i(t)$	The subset of users allocating to robot i
$d(\cdot,\cdot)$	The 2D distance between two points
$D(\cdot,\cdot)$	The 3D distance between two points

to the model. Note that the region $\mathcal{D}$ may enclose some nondeployable zones where UAVs are not allowed to deploy (Fig. 8.1). The UAVs will be deployed at some positions over this area and their altitudes are bounded by the lowest and highest altitudes, denoted by Z_{min} and Z_{max}, respectively. Thus, the deployment space is $\mathcal{S} = \mathcal{D} \times [Z_{min}, Z_{max}]$. A set of users, denoted by U, may stay or move in the area of $\mathcal{D}$. Each user $u \in U$ is associated with the 2D coordinates $u(t) \in \mathcal{D}$ in time slot t. Each UAV i is associated with a 3D position, i.e, $P_i(t) = (x_i(t), y_i(t), z_i(t)) \in \mathcal{S}$. It is also associated with the residual energy $E_i(t)$, with the initial value E_i^0.

When the UAVs move during the monitoring period, it is necessary that the system forms a connected graph (e.g., see Fig. 8.1). Specifically, in such a graph, every UAV needs to have a path linked to one of the base stations. Let $\mathcal{G}$ denote such a graph, which consists of $n + m$ nodes (n UAVs and m base stations). If two UAVs i, j are connected, we have [14]

$$D(P_i(t), P_j(t)) \leq R_1, \tag{8.1}$$

where $D(P_i(t), P_j(t))$ is the standard 3D Euclidean distance between them; and if one UAV i is connected to the base station l, we have

$$D(P_i(t), B_l) \leq R_2, \tag{8.2}$$

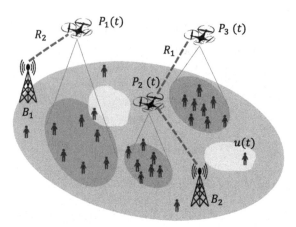

FIGURE 8.1 Illustration of the connected graph with two base stations and three UAVs in an area of interest with two nondeployable zones.

where B_l is the fixed location of the base station l and R_1 and R_2 are computed according to the air-to-air PL and air-to-ground PL as presented in [15].

Furthermore, for the purpose of collision avoidance, the following constraint is considered:

$$D(P_i(t), P_j(t)) \geq R_3, \forall i \neq j, \tag{8.3}$$

i.e., in any slot, the distance between any pair of UAVs i, j should be no smaller than R_3, which is a given constant.

Now, consider the energy consumption model for UAVs. In this chapter, the following power model is used, which is similar to the one used in [11][1]:

$$q_i(t) = \alpha + \beta z_i(t) + \gamma v_i(t), \tag{8.4}$$

where α is the fixed minimum power needed to hover just over the ground, β is a coefficient for the power relating to the altitude [9], $v_i(t)$ is the speed of robot i in time slot t, and γ is the corresponding coefficient. The parameter γ depends on the moving direction of a UAV. For example, moving forward and moving upward vertically correspond to different values of γ. Generally, moving upward consumes more energy than moving forward for the same distance. In this chapter, γ roughly is taken as a constant. The residual energy at UAV i in time slot t can be computed as follows:

$$E_i(t) = E_i^0 - \sum_{t=1}^{t} q_i(t) = E_i^0 - (\alpha t + \beta \sum_{\tau=1}^{t} z_i(\tau) + \gamma s_i(t)), \tag{8.5}$$

where $s_i(t) = \sum_{\tau=1}^{t} v_i(\tau)$ gives the total distance the UAV i has flown. The UAV i still works if $E_i(t) > 0$.

---

[1] Note that the approach developed later is not restricted to this energy consumption model.

Introduce a sign function as follows:

$$\text{Sign1}(x) = \begin{cases} 1, & \text{if } x > \kappa, \\ 0, & \text{otherwise,} \end{cases} \tag{8.6}$$

where κ is a given parameter. When the residual energy drops to κ, a UAV needs to return to the depot. Then, whether UAV i is still working in time slot t can be indicated by $\text{Sign1}(E_i(t))$.

Remark 8.2.1. If a UAV runs out of energy, i.e., $E_i(t) \leq \kappa$, it needs to return to the depot. If it is not a leaf node in the connectivity graph $\mathcal{G}$, it is then replaced by another UAV. This new robot is only used to maintain the connection defined in $\mathcal{G}$; however, the coverage by the new robot is not taken into account in the evaluation.

The served area on the ground of a UAV can be regarded as a disk with radius

$$r_i(t) = f(z_i(t)), \tag{8.7}$$

where $f(\cdot)$ is a given function. If a user is covered by a UAV, the 2D distance between the user and the UAV's projection should be no greater than $r_i(t)$, and the condition of LoS holds. Below, it is always assumed that the UAVs can have LoS with users. This assumption is realistic in some situations such as serving users in an open area like a public square.

Introduce the set $\mathcal{C}(P_1(t), \ldots, P_n(t))$ as the set of all the points $p \in \mathcal{D}$ such that $d(P_i(t), p) \leq \text{Sign1}(E_i(t))r_i(t)$ for some i. Here, $d(P_i(t), p)$ gives the 2D distance between point p and the projection of UAV i on the 2D plane. The set $\mathcal{C}(P_1(t), \ldots, P_n(t))$ is the total coverage area of all the alive UAVs in time slot t. For each user in each time slot, introduce a binary variable $\lambda_u(t)$ to indicate whether it is covered, defined as follows:

$$\lambda_u(t) = \begin{cases} 1, & \text{if } u(t) \in \mathcal{C}(P_1(t), \ldots, P_n(t)), \\ 0, & \text{otherwise.} \end{cases} \tag{8.8}$$

Problem statement: The considered optimization problem is stated as follows. For the given connectivity graph $\mathcal{G}$, the UAV deployment space $\mathcal{S}$, the locations of users $u(t)$, $u \in U$, $t = 1, \ldots, T$, the locations of base stations B_l, $l = 1, \ldots, m$, the initial energy E_i^0, $i = 1, \ldots, n$, and constants $R_1, R_2, R_3, \alpha, \beta, \gamma, \theta$, and κ, find the positions $P_1(t), \ldots, P_n(t) \in \mathcal{S}$ that maximize the total number of covered users in the operation period subject to constraints (8.1), (8.2), and (8.3):

$$\max \sum_{t=1}^{T} \sum_{u \in U} \lambda_u(t) \tag{8.9}$$

$$s.t. \ (8.1), (8.2), (8.3).$$

Before moving to the solution part, the rest of this section analyzes the difficulty of the considered optimization problem. Consider the following relaxations.

- assume that a UAV is connected with a ground base station at any position in $\mathcal{S}$;
- assume that all the UAVs are not constrained by energy capacity;
- assume that the deployment space $\mathcal{S}$ is discretized into grids.

With the first assumption, the connectivity constraints (8.1) and (8.2) are removed. With the second assumption, the objective function (8.9) is not influenced by the energy status of a UAV. With the last assumption, one can have a collection of candidate positions, and each of the candidate positions can cover a subset of users. Furthermore, if the constraint (8.3) is removed, one can obtain a simplified problem: given a set of users, select n positions from the collection such that the number of covered users is maximized. This is the so-called max k cover problem and as shown in [16], it is NP-hard. It is clear that without these relaxations, the original optimization problem is much more complex than the simplified one.

8.3 Proposed solution

In this section, the proposed solution is discussed. An overview of the whole system is provided first, followed by the details of each module.

8.3.1 Overview

The proposed solution consists of several modules, as shown in Fig. 8.2. The location estimation and prediction module is to estimate and predict users' locations based on the photos taken by the camera. The receiver can receive some information from nearby UAVs about users' locations and positions of other robots. The transmitter sends out the position of the UAV as well as some information about users' locations. The flight control module accepts the moving command from MDM and navigates the UAV to the desired position. In this chapter, the focus is MDM. For the others, like communication protocols, channel models, state estimations, and the flight controller, it is simply assumed that they can perform perfectly.

MDM accepts the following input information: its current position, its own covered users' locations, some other users' locations, and nearby robots' positions. Inside MDM, there are several submodules: the residual energy storage, an objective evaluation module, and a connectivity checker. The residual energy storage module keeps track of the residual energy of the UAV according to (8.5). The objective evaluation plays a key role in movement decision by evaluating the objective function. The information taken into account includes the predicted and shared users' locations, the residual energy, and its position. The output is the desired future position. The connectivity checker is embedded with

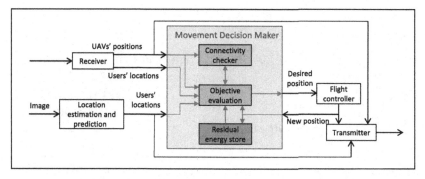

FIGURE 8.2 Structure of UAV controller.

a subgraph of $\mathcal{G}_i$, i.e., it stores the indices of other UAVs or a base station that are connected with itself in $\mathcal{G}$. Then, the connectivity checker checks whether a position can satisfy (8.1) and (8.2). The constraint (8.3) is also checked here. If all the results are positive, the checked position is further evaluated by the objective evaluation module; otherwise, such a position is abandoned by the objective evaluation module.

8.3.2 Supporting modules

Before introducing the main module, the details of the supporting modules are first discussed.

The estimation and prediction module is to estimate and predict the users' locations, which are seen by the onboard camera of a UAV [12]. There are many approaches in the literature for user detection and movement prediction. Since they are beyond the scope of this chapter, it is simply assumed that the robot can recognize all the seen users and predict the future movements based on some well-studied mobility models [17]. It is worth mentioning that in practice the estimation and prediction module may not always work perfectly. In other words, there may always exist some errors in the estimations and predictions. The errors may make the actual coverage of users far from the expectation. In Section 8.4, we will present some simulations to demonstrate the sensitivity to the prediction error. Another point is that to reduce the impact of the errors on the coverage performance, it is necessary to make the UAVs react fast to the predictions and the predictions can only be used in the next time slot.

The mobility of users is not controlled by the system; thus, it is possible that some future locations of users, which are currently covered by the UAV, may be outside the coverage range of the UAV in the future. With this regard, the users can be grouped into two categories based on their predicted locations: in-range and out-of-range users. Specifically, the in-range users include those which are more likely to be covered by the UAV in the future, while the out-of-range users include those which are more likely to be covered by others.

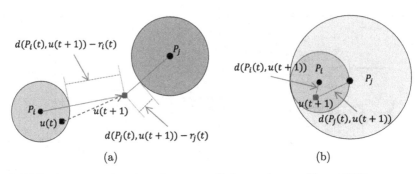

FIGURE 8.3   (a) Illustration of out-of-range user. (b) One user is covered by two UAVs.

The type of users can be determined by comparing the distances between a predicted location and the covering circles of UAVs. Take Fig. 8.3a as an example. There are two UAVs i, j deployed at $P_i(t)$, $P_j(t)$ and their coverage radii are $r_i(t)$ and $r_j(t)$, respectively. Consider a user u which is covered by UAV i in time slot t (see the black square). The predicted location in the next slot is shown by the red square (dark gray in print version). The 2D distance between user u and the nearest point on a covering circle of UAV i is given by $d(P_i(t), u(t+1)) - r_i(t)$, and that for UAV j is given by $d(P_j(t), u(t+1)) - r_j(t)$. If $d(P_i(t), u(t+1)) - r_i(t) < d(P_j(t), u(t+1)) - r_j(t)$ $\forall j \neq i$, u is in range to UAV i; if $d(P_i(t), u(t+1)) - r_i(t) > d(P_j(t), u(t+1)) - r_j(t)$ $\exists j \neq i$, this user is out of range to UAV i and it is shared with UAV j, and if $d(P_i(t), u(t+1)) - r_i(t) = d(P_j(t), u(t+1)) - r_j(t)$, UAV i can share it or not randomly. In contrast, if one user's future position is within the vision cones of two or more UAVs, it is said to be covered by the closest UAV. In the example shown in Fig. 8.3b, the user u is said to be covered by UAV i rather than j, since $d(P_i(t), u(t+1)) < d(P_j(t), u(t+1))$.

The out-of-range users' locations will be sent to other nearby UAVs by the transmitter module. The transmitter takes the positions of nearby UAVs into account and correspondingly sends each user's location to the nearest robot. Also, the new position of the UAV is sent to the nearby UAVs for collaboration. It is worth pointing out that the union of all the in-range and out-of-range users may be only a subset of U, since there may exist some users which cannot be seen by any camera.

Regarding the receiver of a UAV, say i, it receives the positions of other nearby UAVs, say j, and the shared users' locations. These shared users' locations are promising to be covered by UAV i from the point of UAV j. Thus, the in-range users and the users introduced by other UAVs are the potential users to be covered by a UAV.

The last module is the flight controller. As mentioned above, this module accepts the desired destination from MDM and we assume that the flight controller is capable to drive the UAV there. Note that the collision-free constraint

(8.3) only guarantees that each final destination is collision-free. Thus, the flight controller also needs to generate a collision-free path to that destination.

8.3.3 Movement decision maker

Taking into account the in-range users and the users introduced by other UAVs, the current position, and the residual energy, the task of the MDM is to find a new position such that it can make the largest contribution to the objective function (8.9) without disobeying constraints (8.1), (8.2), and (8.3). Note that the objective function (8.9) is in a centralized form. To develop a decentralized navigation algorithm, the decentralized form is required.

Introduce another sign function as follows:

$$\text{Sign2}(x) = \begin{cases} 1, & \text{if } x \geq 0, \\ 0, & \text{otherwise.} \end{cases} \tag{8.10}$$

Then, whether user u is covered by UAV i in time slot t can be indicated by $\text{Sign2}(r_i(t) - d(P_i(t), u(t)))$, where $d(P_i(t), u(t))$ gives the 2D distance between the projection of robot i and user u in time slot t.

Let $U_i(t)$ be the set of users' locations stored by the UAV i in time slot t, including the in-range users as well as the shared users. As discussed in Section 8.3.2, one user can only exist in at most one coverage set. Thus, for any two UAVs, their coverage sets are always disjoint, i.e., $U_i(t) \bigcap U_j(t) = \emptyset$ if $i \neq j$. With (8.10), the number of covered users by the UAV i is computed by

$$N_i(t) = \sum_{u(t) \in U_i(t)} \text{Sign1}(E_i(t))\text{Sign2}(r_i(t) - d(P_i(t), u(t))). \tag{8.11}$$

Furthermore, the total number of covered users by the UAV network during the monitoring period is given by

$$\sum_{t=1}^{T} \sum_{i=1}^{n} N_i(t). \tag{8.12}$$

Now, the objective function is in a decentralized form.

For a UAV i, it cannot change the coverage of users in the previous slots, but it can try to maximize its contribution in the remaining time slots. It is clear that the overall objective of the system coverage performance depends on two factors: (1) whether a UAV can still work and (2) if so, how many users it can cover. According to the considered energy consumption power model and the coverage model, these two objectives are contradictory. More specifically, to increase the number of covered users, a UAV needs to enlarge the coverage area, in which case it has to be at a higher altitude; at the same time, the energy consumption power increases as well, which decreases the working time. Therefore, a balance between them needs to be explored. The proposed solution is to replace (8.12) with one which maximizes the product of the working time and the number of covered users.

The UAV i, which is at $P_i(t)$ with the residual energy $E_i(t)$ in time slot t, needs to decide the new position $P_i(t+1)$ for the time slot $t+1$. After receiving the shared information from the nearby UAVs, the robot i knows the set $U_i(t+1)$. The number of covered users depends on the distribution of users in $U_i(t+1)$ and $P_i(t+1)$. Since $U_i(t+1)$ is given, the number of covered users in time slot $t+1$ can be written as a function of the position $P_i(t+1)$, i.e., $N_i(t+1) = f(P_i(t+1))$. Moreover, the remaining working time is estimated by $\frac{E_i(t+1)}{\alpha+\beta z_i(t+1)}$, under the assumption that the UAV i will stay at the altitude $z_i(t+1)$, where $E_i(t+1)$ gives the residual energy after reaching the new position in the time slot $t+1$. Obviously, $E_i(t+1)$ depends on $P_i(t+1)$. Considering that the left monitoring time is $\mathcal{T} - t\frac{\mathcal{T}}{T}$, if $\frac{E_i(t+1)}{\alpha+\beta z_i(t+1)} > \mathcal{T} - t\frac{\mathcal{T}}{T}$, which means the current residual energy is sufficient for the UAV i to work for all the rest of the monitoring time, the UAV will only work for $\mathcal{T} - t\frac{\mathcal{T}}{T}$. Thus, the residual working time of UAV i is given by

$$g(P_i(t+1)) \triangleq \min\{\frac{E_i(t+1)}{\alpha+\beta z_i(t+1)}, \mathcal{T} - t\frac{\mathcal{T}}{T}\}. \tag{8.13}$$

The expected contribution made by the UAV i from time slot $t+1$ to the end of operation is estimated by

$$\max_{P_i(t+1)} g(P_i(t+1))f(P_i(t+1)). \tag{8.14}$$

Now, we present the method to solve the problem (8.14), (8.1), (8.2), (8.3) (see Algorithm 1). This algorithm navigates each UAV individually. For the UAV i, a candidate horizontal position is the middle point of the line segment connecting two users in $U_i(t+1)$ (lines 3 and 4). It then computes the altitude corresponding to the radius (half of the line segment, line 5) by (8.7) and checks whether such position is reachable. Since a UAV needs to fly to a new position in a limited time and it is constrained by speed, the set $\mathcal{P}_i(t)$ of all the positions a UAV can reach from its current position $P_i(t)$ is introduced. The shape of the set $\mathcal{P}_i(t)$ depends on the possible speeds in different directions. Suppose the UAV can maintain the same maximum speed in all directions. Such set is a sphere centered at $P_i(t)$. However, in most practical cases, the speed for moving upward may be smaller than those of other directions. Note that since the region $\mathcal{D}$ may enclose some nondeployable zones, for any UAV, the set $P_i(t)$ does not contain any points whose projections on the ground are inside these zones. If a candidate position is in $\mathcal{P}_i(t)$, it then checks whether it is within the range of $[Z_{min}, Z_{max}]$. If so, it further checks whether this 3D position satisfies (8.1), (8.2), and (8.3). If so again, it evaluates the position by computing the objective function (8.14). After evaluating all the candidates, the position which maximizes (8.14) can be selected.

Algorithm 1 The proposed algorithm running in MDM.

1: **for** Any user pair u and w in $U_i(t+1)$ $(u \neq w)$ **do**
2:     $h_i(u, w) \leftarrow 0.$
3:     $x_i(u, w) \leftarrow \frac{u^x(t+1)+v^x(t+1)}{2}.$
4:     $y_i(u, w) \leftarrow \frac{u^y(t+1)+v^y(t+1)}{2}.$
5:     $r_i(u, w) \leftarrow \frac{d(u(t+1),v(t+1))}{2}.$
6:     $z_i(u, w) \leftarrow \frac{r_i(u,w)}{\tan(\frac{\theta}{2})}.$
7:     **if** $z_i(u, w) \in [Z_{min}, Z_{max}]$ & $P_i(u, w) \in \mathcal{P}_i(t)$ & $P_i(u, w)$ satisfies
        (8.1), (8.2), (8.3) **then**
8:         Compute objective value $h_i(u, w)$ by (8.14).
9:     **end if**
10: **end for**
11: Find P_i corresponding to the largest h_i.

Remark 8.3.1. An implied assumption in Algorithm 1 is that there are at least two users in the set $U_i(t+1)$, i.e., $|U_i(t+1)| \geq 2$. If $|U_i(t+1)| = 1$, the UAV will work just over the user at Z_{min} if this position is reachable. If $|U_i(t+1)| = 0$, the UAV randomly chooses a moving direction. Thus, the proposed method can find users that are not seen by the UAVs only in a random manner, which is also one demerit.

Remark 8.3.2. For an individual UAV, it only needs to consider the users that are already covered by itself and those introduced from other nearby UAVs. With the union of the two groups of users, the UAV computes the best position which achieves the largest coverage, with the consideration of its residual working time. At any position, when the users' locations are given, a UAV knows how many users are covered, based on the coverage model.

To execute Algorithm 1, the robot will try $C^2_{|U_i(t+1)|}$ combinations to figure out the best 3D position, i.e., selecting two users from $|U_i(t+1)|$ users. Then, the time complexity is $O(|U_i(t+1)|^2)$. For each combination, it needs to check how many of the remaining $|U_i(t+1)| - 2$ users are within the coverage area. Thus, the overall time complexity is $O(|U_i(t+1)|^3)$. Note that since the users are mobile, this approach is not guaranteed to keep achieving a larger coverage contribution. But, for the case with static users, it is easy to see that the proposed approach navigates the UAVs from initial positions to better positions in terms of the product of the working time and the number of covered users.

So far, how a single UAV navigates itself based on the measured and shared information to maximize its own coverage contribution for the network has been discussed. This section is ended by presenting how the whole system works.

Considering the fact that the operation of each UAV depends on the shared information by the nearby UAVs, in this approach, the UAVs operate in a serial manner. In particular, each time slot is further divided into n subslots. All the

UAVs are with the same clock. Each UAV has the unique label $i \in [1, n]$. In the ith subslot of a time slot, the UAV i takes actions while all the others hover at their current positions. During this subslot, robot i broadcasts information including its position as well as the out-of-range users via its transmitter. At the same time, all the other robots turn off their transmitters but turn on their receivers. It is easy to understand that it is not necessary for all the UAVs to hear a broadcast. because for a UAV which is far from the sender, the broadcast information does not impact the former's movement. Therefore, a sender only needs to set a transmitting power such that the UAVs within a certain range can hear it. When the ith subslot expires, the robot i stops moving, and the robot $i + 1$ starts to move if $i + 1 \leq n$. If $i + 1 > n$, the UAV labeled 1 will start to move. This process will repeat for the whole period of time. To sum up, the UAV network will work as follows:

A1: The UAVs start with some initial positions, such that they form a connected graph with base stations, i.e., any UAV has a valid path to at least one base station.

A2: For any slot $t = 1, \ldots, T$ and any subslot $i = 1, \ldots, n$, if the UAV i is alive, it computes its future position by Algorithm 1 and moves there; if $E_i(t) \leq \kappa$ and the robot i is not a leaf node in $\mathcal{G}$, it is replaced by a new robot and the new robot remains at the position where robot i exhausts its energy.

8.4 Simulation results

In this section, the performance of the proposed algorithm is evaluated using MATLAB®.

We consider an area with a size of 1200 m by 800 m (Fig. 8.4). The deployment region $\mathcal{D}$ is bounded by the black solid line and it has two holes where the UAVs are not allowed to enter. There are three base stations marked by a black ×. The system parameters are specified as follows: $Z_{min} = 20$ m, $Z_{max} = 100$ m, $R_1 = 300$ m, $R_2 = 200$ m, $R_3 = 20$ m, $\alpha = 1$, $\beta = 0.1$, $\gamma = 0.01$, $E^0 = 100$, $\kappa = 5$, and $\theta = 120°$. Also, $\mathcal{T} = 20$ minutes and $T = 20$. The speed of the UAVs is set as 10 m/s.

We first consider a simple case with 200 static users to make the results tractable. The users are randomly deployed, and some of them may be outside $\mathcal{D}$ (Fig. 8.5a). We consider 10 UAVs in this case, whose initial positions in the horizontal plane are shown in Fig. 8.5a. The connectivity graph between the UAVs and the base stations is also presented in Fig. 8.5a. The robots are all at an altitude of 60 m. By applying the proposed method, the horizontal and vertical movements of the UAVs are shown in Fig. 8.5b and Fig. 8.5c, respectively. From Fig. 8.5b we can see that no UAVs enter the nondeployable zones. When a UAV exhausts the energy, its altitude is set as zero in Fig. 8.5c. We can see that the first UAVs run out of energy in the 17th slot, and at the end of the operation period, only two robots are still alive. The number of covered users in each

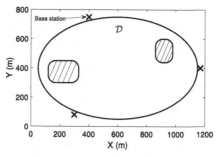

FIGURE 8.4 The deployment region D bounded by the black solid line with two holes. There are three base stations outside D marked by a black ×.

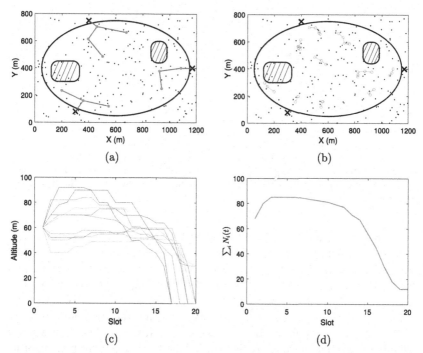

FIGURE 8.5 Simulation results for a simple case with static users. (a) The distribution of users and the connection graph. (b) The horizontal movements of the UAVs. (c) The vertical movements of the UAVs. (d) The number of covered users in each time slot by the alive UAVs.

slot is shown in Fig. 8.5d. From Figs. 8.5b and 8.5d we can see that the UAVs generally decrease their altitudes in the second half of the operation period. Accordingly, the number of covered users decreases significantly. The reason is that with the decrease of the residual energy, the robots reduce their altitudes such that they can serve for a longer time. Clearly, the objective value is the accumulated value of the figures in Fig. 8.5d.

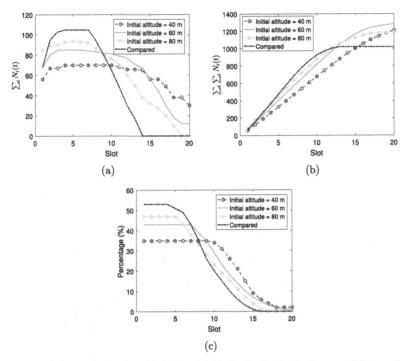

FIGURE 8.6 Simulation results with different initial altitudes for the simple case. (a) The number of covered users in each slot. (b) The accumulated number of covered users. (c) How long a user has been monitored.

Moreover, we consider the influence of the initial altitudes. With the same initial positions in the horizontal plane as the above case, we further conduct simulations with altitudes of 40 m and 80 m. The numbers of covered users for these cases are summarized in Fig. 8.6. From Fig. 8.6a we can see that when the UAVs are at a higher altitude (80 m), more users can be seen by them initially; however, more energy is consumed, which results in more robots exhausting their energy faster. By contrast, when the robots start from a lower altitude (40 m), the coverage is smaller, but the lifetime is longer. Comparing these three sets of simulations, the one with the initial altitude of 60 m performs best in terms of the considered objective (see Fig. 8.6b). Furthermore, we present details about the percentages of users being covered for the different number of slots during the operation period. Specifically, we demonstrate the number of slots each user has been covered. As shown in Fig. 8.6c, 34% of the users have been covered for 10 slots for the case with an initial altitude of 40 m, and the percentages for the cases with initial altitudes of 60 m and 80 m are 29% and 23%, respectively. For the case of an initial altitude of 40 m, 2% of the users have been covered for the whole operation period. The figures for the other two cases are 1% and 0%, respectively.

Now, we compare the proposed approach with the approach proposed in [4] (called the compared approach in the remainder of the chapter). The compared approach considers the reactive deployment of UAVs in 3D for surveillance purposes. The compared approach targets coverage only while it does take the limited energy constraint into account, which is the main difference from the current chapter. Another difference is that the compared approach specifically focuses on the quality of coverage, which is influenced by not only the distance between a user and its associated robot but also the user's position in the vision cone. In contrast, in the current approach, if a user is within the vision cone, it is regarded as being covered. The third difference is that the compared approach navigates each UAV only according to the users it can see, while the proposed approach further accounts for the users that are shared by other nearby UAVs. For the above simple case, the compared approach is applied and the initial positions of 10 UAVs and the connectivity graph are the same as in the simulation with the initial altitude of 60 m. The simulation results are presented in Fig. 8.6 as well. By observing the red and black curves we can see that the compared approach achieves a larger number of covered users at the beginning (Fig. 8.6a). However, since it does not take the energy capacity limitation into account, the UAVs generally fly to a higher altitude for a better quality of coverage. Thus, the UAVs exhaust their energy faster than those navigated by the proposed approach. The overall performance in terms of coverage of the proposed approach is better than that of the compared one by 26% (Fig. 8.6b). Regarding how long a user has been monitored by the compared approach, Fig. 8.6c shows that only 20% of users are covered for 10 slots and no users have been covered for more than 16 slots. Note that it is worth comparing with the optimal solution. However, due to the complexity of the considered problem, we have not found a good way to compute the optimal solution.

In the above cases, we consider static users and the simulations are run only one time. Below, we consider cases with mobile users and the simulations are run multiple times. Thus, the results we present are statistic values. In the below simulations, all users are randomly deployed and each user can be static or choose a random moving direction and move for a random distance. The speed of the users is upper-bounded by 1 m/s. If a user meets the boundary of the area, rather than leaving the area, it makes a turn to move back to the area. For a fair comparison, in each case, the UAVs are with the same initial positions and the connectivity graphs are the same as well in the two approaches. Note that as the users are mobile, the optimal solution cannot be obtained using brute-force search. So below we only present the results of the proposed approach and the compared approach.

Besides 10 UAVs, we also conduct simulations with other numbers of UAVs. As the users are randomly mobile, to mitigate the randomness, all results have been averaged over 20 independent runs of simulations. Fig. 8.7a demonstrates the average coverage by one UAV, which is represented in terms of ratio. With an

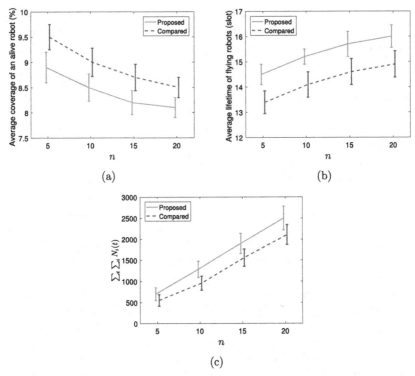

FIGURE 8.7 Simulation results with mobile users for different numbers of UAVs. (a) Average coverage of UAVs. (b) Average lifetime of UAVs. (c) The overall coverage.

increasing number of UAVs n, the average coverage shows a decreasing trend. The reason is that when fewer UAVs are available, the distance between a pair of UAVs is larger than the case when more UAVs are used. Thus, each robot has the potential to cover more users by changing its position. However, due to the same reason, the movement of an individual robot may be more frequent and the moving distance may be longer when fewer robots are in use than the case with more robots. Thus, for small n, the average lifetime of a robot is shorter in than the case with large n (Fig. 8.7b). Comparing the two methods, the compared one generally can see more users due to its optimizing metric (quality of monitoring) (Fig. 8.7a), while a UAV in the compared method often serves at a higher altitude for a better quality of monitoring. Thus, the average lifetime of a robot is shorter than that in the proposed method (Fig. 8.7b). Regarding the overall coverage of users, with increasing n, both methods show an increasing trend (Fig. 8.7c). The proposed method outperforms the compared one because the gap in the average lifetime of a robot is larger than the gap in the average coverage of a robot. Through these simulations, we can see the proposed approach performs better than the compared one.

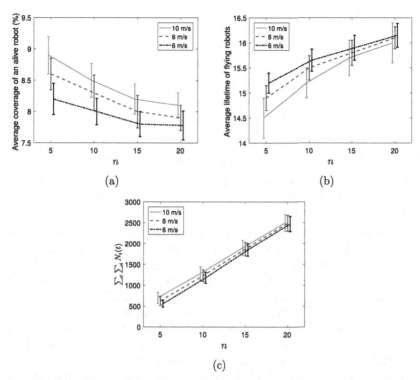

FIGURE 8.8 Simulation results with mobile users for speeds. (a) Average coverage of UAVs. (b) Average lifetime of UAVs. (c) The overall coverage.

We also studied the impact of the robot speed on coverage performance. In the above simulations, the speed of robots is set as 10 m/s. Now, we consider some other values for the speed. Again, we conduct simulations with mobile users and demonstrate the average results below. As seen in Fig. 8.8a, for a certain number of UAVs, the larger the speed, the more users one robot can cover on average. The reason is that with a larger speed, a robot can move to a new position within a larger space, which increases the possibility of finding a better position in terms of coverage. However, at the same time, this also leads to more energy consumption due to the long-distance movement (Fig. 8.8b). With increasing n, such influence becomes smaller (Figs. 8.8a and 8.8b). The reason is that with more UAVs in use, the coverage of each robot becomes smaller. Although the feasible movement space becomes larger when the speed is higher, it does not improve the coverage significantly (Fig. 8.8a). Thus, the average lifetimes of UAVs in cases with different speeds become close (Fig. 8.8b). The impact of speed on the performance in terms of overall coverage also becomes smaller when the number of robots increases (Fig. 8.8c).

The above simulations with mobile users assume that the predictions of the users' movements are accurate, which is not realistic in practice. To end this

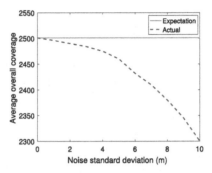

FIGURE 8.9 The influence of the inaccurate prediction of user movements on the coverage performance with 20 UAVs. The expectation value is the average result shown in Fig. 8.7c.

section, we investigate the influence of uncertainty in the prediction. Specifically, we add some noise to the users' movements. The noise is with zero mean value and the standard deviation indicates the accuracy of the prediction (smaller standard deviation indicates higher prediction accuracy). The results of the noise-free case represent the expectation, while those of the noisy case give the actual performance. The results are shown in Fig. 8.9 (mean values only). Clearly, when the standard deviation is small, e.g., smaller than 3 m, the actual performance is very close to the expectation. The reason is that with a small standard deviation, only those users which are near the coverage circle of UAVs are influenced, while the inner users that are far from coverage circles are not impacted much. When they do not move towards the predicted directions, the UAVs may lose the coverage of a small fraction of users. However, with increasing standard deviation, the gap between actual coverage and expectation becomes larger, because both the users near the coverage circles and the inner users may disobey the predicted movements. This leads to a large gap between the actual coverage and the expected value.

8.5 Conclusion

This chapter considered the problem of deploying a fleet of UAVs to monitor users on a 2D plane. To improve the service quality, the problem of maximizing the operating time of UAVs together with the number of covered users was studied. A decentralized algorithm to navigate UAVs in 3D space was developed. This algorithm is easy to implement in real-time. The comparisons with a previously published approach indicated that the proposed method achieves better network lifetime and user coverage. One future research direction is to extend the current method to the scenario where users are on 3D terrains, which is more practical but also challenging. Such an extension requires at least a new coverage model to enclose the blockage caused by some high-rise buildings. Moreover, another important direction of future research will be to obtain theoretical performance guarantees for the proposed reactive deployment algorithm.

References

[1] P. Mekikis, A. Antonopoulos, E. Kartsakli, L. Alonso, C. Verikoukis, Communication recovery with emergency aerial networks, IEEE Transactions on Consumer Electronics 63 (August 2017) 291–299.

[2] S.A. Hadiwardoyo, E. Hernández-Orallo, C.T. Calafate, J.C. Cano, P. Manzoni, Experimental characterization of UAV-to-car communications, Computer Networks 136 (2018) 105–118.

[3] H. Huang, A.V. Savkin, M. Ding, C. Huang, Mobile robots in wireless sensor networks: a survey on tasks, Computer Networks 148 (2019) 1–19.

[4] H. Huang, A.V. Savkin, An algorithm of reactive collision free 3-D deployment of networked unmanned aerial vehicles for surveillance and monitoring, IEEE Transactions on Industrial Informatics 16 (1) (2020) 132–140.

[5] E. Yanmaz, S. Yahyanejad, B. Rinner, H. Hellwagner, C. Bettstetter, Drone networks: communications, coordination, and sensing, Ad Hoc Networks 68 (2018) 1–15.

[6] A.V. Savkin, H. Huang, A method for optimized deployment of a network of surveillance aerial drones, IEEE Systems Journal 13 (4) (2019) 4474–4477.

[7] M. Erdelj, M. Król, E. Natalizio, Wireless sensor networks and multi-UAV systems for natural disaster management, Computer Networks 124 (2017) 72–86.

[8] C. Caillouet, F. Giroire, T. Razafindralambo, Optimization of mobile sensor coverage with UAVs, in: Conference on Computer Communications Workshops (INFOCOM WKSHPS), IEEE, 2018, pp. 622–627.

[9] L.D.P. Pugliese, F. Guerriero, D. Zorbas, T. Razafindralambo, Modelling the mobile target covering problem using flying drones, Optimization Letters 10 (5) (2016) 1021–1052.

[10] J. Lyu, Y. Zeng, R. Zhang, T.J. Lim, Placement optimization of UAV-mounted mobile base stations, IEEE Communications Letters 21 (March 2017) 604–607.

[11] D. Zorbas, L.D.P. Pugliese, T. Razafindralambo, F. Guerriero, Optimal drone placement and cost-efficient target coverage, Journal of Network and Computer Applications 75 (2016) 16–31.

[12] H. Zhao, H. Wang, W. Wu, J. Wei, Deployment algorithms for UAV airborne networks towards on-demand coverage, IEEE Journal on Selected Areas in Communications 36 (9) (2018) 2015–2031.

[13] H. Huang, A.V. Savkin, Reactive 3D deployment of a flying robotic network for surveillance of mobile targets, Computer Networks 161 (2019) 172–182.

[14] A.V. Savkin, H. Huang, Deployment of unmanned aerial vehicle base stations for optimal quality of coverage, IEEE Wireless Communications Letters 8 (1) (2019) 321–324.

[15] A. Al-Hourani, S. Kandeepan, S. Lardner, Optimal LAP altitude for maximum coverage, IEEE Wireless Communications Letters 3 (Dec 2014) 569–572.

[16] U. Feige, A threshold of ln n for approximating set cover, Journal of the ACM (JACM) 45 (4) (1998) 634–652.

[17] M. Khan, K. Heurtefeux, A. Mohamed, K.A. Harras, M.M. Hassan, Mobile target coverage and tracking on drone-be-gone UAV cyber-physical testbed, IEEE Systems Journal 12 (4) (2018) 3485–3496.

Chapter 9

Optimized deployment of UAV base stations for providing wireless communication service in urban environments[☆]

9.1 Motivation

Explosive demands for mobile data are driving mobile operators to respond to the challenging requirements of higher capacity and improved user experience [1,2]. Deploying more SBSs can meet the increasing traffic demand. This solution, however, may result in more cost for site rental and bring along other issues, such as a high percentage of SBSs having low utility in nonpeak hours. In this context, the utilization of UAV-BSs could be a more efficient solution than SBS densification.

Recently, great effort has been put into using UAV-BSs to assist wireless communication systems. Some publications focus on the empirical and analytical studies on air-to-ground channel modeling. In [3], the air-to-ground PL is modeled. It shows the situations of LoS between a UAV-BS and a ground user and NLoS, i.e., the user still receives a signal from the UAV-BS due to strong reflections and diffraction. A closed-form expression for the probability of LoS between a UAV and a user is developed in [4], and then the optimum altitude that maximizes the radio coverage is obtained via an analytical method. Furthermore, the authors of [5] consider the deterministic ITU channel model [6] by utilizing more information about the city map, such as the shapes and heights of buildings.

Like the conventional system using SBSs, planning and optimization of UAV-BSs are still fundamental tasks. Generally speaking, they refer to determining the number and locations of UAV-BSs to provide wireless access to users [7]. The determination of the number of UAV-BSs aims at providing the required capacity for the expected demand volume, and the determination of the positions of UAV-BSs aims at providing a certain targeted quality of service.

☆ The main results of the chapter were originally published in Hailong Huang, Andrey V. Savkin, Ming Ding, Mohamed Ali Kaafar, Optimized deployment of UAV base station to improve user experience in cellular networks, Journal of Network and Computer Applications 144 (2019) 49–58. Permission from Elsevier for reuse was obtained.

There is a growing number of publications on the topic of UAV-BS placement. Reference [8] uses UAV-BSs as relays to offload traffic of SBSs and discusses the optimal position of a single UAV-BS for maximizing the data rate between an SBS and a user. Reference [9] also considers the UAV-BS relay application. The objective is to find suitable positions for UAV-BSs that maximize the downlink throughput. Reference [10] studies the problem of user demand-based UAV-BS assignment over geographical areas subject to traffic demands. Reference [11] considers the issue of the minimum number of UAV-BSs and the positions to cover all the users, and the paper [12] considers a similar problem. The authors of [13] focus on the problem of deploying the minimal number of UAVs at the same altitude to completely cover a given area of interest, and the publication [14] further addresses a similar problem but the UAVs can be deployed at different altitudes. Another group of publications focuses on the maximum coverage by a given number of UAV-BSs. The paper [15] proposes a decentralized method to maximize the coverage of users and minimize the communication cost between UAV-BSs with the assumption that users are located on some street graph. The paper [16] presents a distributed algorithm to find locally optimal positions for a set of UAVs. The reference [17] presents a K-means algorithm to cluster a given set of users considering the capacity of UAV-BSs and the uncovered users are served by the background SBS. The publication [18] optimizes the positions of UAV-BSs based on a Voronoi cell model.

One disadvantage in most of the aforementioned references is the UAV-BSs are deployed at the same altitude, which does not explore the flexibility of the vertical dimension and thus are 2D problems in essence. There are also some methods in the literature focusing on the 3D scenario. Reference [19] considers the problem of deploying one UAV-BS in a 3D environment to maximize the number of served users. Exploring the relationship between vertical and horizontal dimensions, reference [20] solves the problem in [19] by turning the problem into a circle placement problem. Reference [21] also considers the issue of the minimum number of UAV-BSs and their deployments like [11]. The altitude is taken as a new variable and a PSO-based heuristic algorithm is proposed. The reference [22] focuses on characterizing the quality of a coverage model, based on which a decentralized navigation algorithm is proposed to guide each UAV individually in the horizontal direction and vertical direction, respectively. Though adding one more dimension makes the problem even more complex, the benefit will be the increased system performance because the UAV-BSs can adjust their altitudes to reduce the interference to users which are covered by nearby UAV-BSs.

Although UAV-BSs are promising to improve the user experience in wireless networks, several issues have not been comprehensively considered in the existing approaches. The first one is the energy constraint of UAV-BSs. The state-of-the-art UAV-BSs are usually powered by batteries, which are limited in practice due to the size and weight constraints of UAV-BSs. In this regard, a UAV-BS management method is necessary to ensure seamless coverage of the given area. Another limitation is that most approaches assume the simplified channel models based on the average characteristics of the environment. However, they may not achieve the expected gains with the generated UAV-BS deployments in a particular urban environment setting. In this chapter, we focus on the 3D UAV-BS deployment problem, which is to find the optimal 3D positions for UAV-BSs to serve ground users on a 2D plane. The positions of UAV-BSs are crucial to the user experience. For one thing, in the urban environment, due to the existence of dense buildings, a user may have LoS or NLoS links with a UAV-BS, which significantly influences the received power at the user side. For the other, some other UAV-BSs rather than the serving one may introduce interference to the user, which further reduces the SINR and then the spectral efficiency (SE). Thus, the deployment of UAV-BSs should take into account the coverage of users and the user experience simultaneously. For the energy constraint issue, we consider that a set of charging platforms or autonomous battery swap platforms [23,24] is available (below we use the term charging station). Then, the routine of a UAV-BS is serving users, flying to a charging platform, recharging on it, and then flying back to its serving position. According to the field experiments using Phantom UAV-BSs [25], the power for flying is over 140 W, while the typical power for transmitting information through radios is usually around 1 W [26], which is much lower than the former case. Compared to the physical movements, the energy consumption caused by wireless transmission is neglected. To guarantee a certain time for serving users, the positions of UAV-BSs should be well managed such that they are able to get recharged before running out of battery.

We assume that the geographic information of the area of interest is known. It includes the sizes, heights, and distribution of buildings, which can be easily measured. We assume that only the outdoor users are the target users, while indoor users can be served by indoor SBSs. We also assume that the indoor-to-outdoor penetration loss is high enough such that an outdoor user cannot receive any signal or interference from indoor SBSs [27]. Furthermore, we assume that the UAV-BSs are wirelessly connected to some outdoor SBSs via high-frequency radios, which do not interfere with the low-frequency radios between UAV-BSs and users. For any location on the ground, we associate it with a user density function to indicate the number of users at this location and a height function to indicate the height of the building upon it.

We formulate an optimization problem to find the optimal 3D positions for a given number of UAV-BSs, with the objective of maximizing the number of covered users, subject to the constraints that (1) all the UAV-BSs are positioned at safe locations (which will be defined later); (2) all the UAV-BSs are able to fly to the nearest charging station for charging the batteries and guarantee a certain serving time; and (3) the SINR at any covered user should be above the given threshold. Furthermore, we show that this problem is NP-hard. Then, a greedy algorithm is proposed together with the computational complexity analysis. We conduct extensive computer simulations to illustrate the effectiveness of the proposed algorithm. Comparing to the existing approaches like [10,11,15,17,18], this paper deploys UAV-BSs in 3D space, which can achieve better performance in terms of interference. Comparing to some 3D approaches including [19–21], this paper adopts the deterministic PL model by taking the environmental factors into consideration, which is more precise than the statistical model. Different from [19,20], which focus on a single-UAV-BS case, this chapter considers the deployment of multiple UAV-BSs; and different from [21], which tries to find the number of UAV-BSs from the viewpoint of system plan, this chapter considers how to deploy a given set of UAV-BSs. The main results of the chapter were originally published in [28].

The remainder of the chapter is organized as follows. In Section 9.2, we present the system model and state the considered problem formally. In Section 9.3, we analyze the difficulty of the studied problem and present the proposed solutions together with the time complexity analysis. In Section 9.4, we show extensive simulations to evaluate the proposed approach. Finally, Section 9.5 summarizes this chapter and discusses future work.

9.2 System model and problem statement

This section presents the system model and states the studied problem.

Consider an urban area of interest to deploy UAV-BSs to serve ground outdoor users. We first assume that the map of this area is available. In particular, we use V to capture the information on the map; V is a 3D space model and the ranges of the three dimensions are $[X_{min}, X_{max}]$, $[Y_{min}, Y_{max}]$, and $[0, H_{max}]$, respectively. Let the ground map be S. A point $p \in S$ has the plane coordinates (X, Y), and $X \in [X_{min}, X_{max}]$, $Y \in [Y_{min}, Y_{max}]$. Let $h(p)$ or $h(X, Y)$ denote the height of building at (X, Y). Further, a point $p \in V$ has the 3D coordinates (X, Y, Z) and $Z \in [0, H_{max}]$. A point $(X, Y, Z) \in V$ is said to be outside a building if $h(X, Y) < Z$. The area of interest may also contain some NFZs, which can be considered as buildings in the current model, and the corresponding height is H_{max}.

Consider that there is a set of UAV-BSs $\mathcal{N}$ serving in the field and the number of UAV-BSs is $n = |\mathcal{N}|$. UAV i ($i \in \mathcal{N}$) has the 3D coordinates $P_i = (x_i, y_i, z_i) \in V$. Basically, the position of a UAV-BS must be within V and the altitude should not be lower than H_{min}. Let $d(P_i, p)$ be the Euclidean distance between UAV-BS i at $P_i = (x_i, y_i, z_i)$ and a point at $p = (X, Y, Z) \in V$, i.e.,

$$d(P_i, p) = \sqrt{(x_i - X)^2 + (y_i - Y)^2 + (z_i - Z)^2}. \tag{9.1}$$

In the urban environment, the security issue related to UAV-BSs needs to be considered. We introduce a constant $H_s < H_{min}$ to represent the safety distance. We require that all the UAV-BSs must be at least H_s away from buildings. Let U_i be a set of points of V which are within the distance H_s to UAV-BS i. Then, $U_i = \{p | d(P_i, p) < H_s, p \in V\}$ and U_i is a ball. We say that the position P_i is safe if

$$h(X, Y) \leq Z, \ \forall \ (X, Y, Z) \in U_i. \tag{9.2}$$

In other words, constraint (9.2) says that if P_i is safe, any point in the set U_i should be outside buildings.

We introduce a virtual map G_i for UAV-BS i, which is parallel to S and consists of the points of S whose heights are smaller than $z_i - H_s$, i.e., $G_i = \{p|h(p) < z_i - H_s, p \in S\}$. It is easy to understand that UAV-BS i can reach any point on G_i without changing its altitude. Such a virtual map will be used in the discussion of the trajectory of UAV-BSs when they fly between serving positions and charging stations below.

Now we consider the energy constraint of UAV-BSs. Since the currently available commercial UAV-BSs are usually powered by preloaded batteries, their operation time is limited. If the overall serving time of UAV-BSs in the area is several orders of the operation time of a single battery, to maintain a seamless coverage, we need to consider the charging issues. We assume the UAV-BSs can recharge their batteries on the charging stations existing in the field. Suppose that there is a set of charging stations $\mathcal{M}$ located at some fixed points $Q_j \in S$, $j \in \mathcal{M}$. To simplify the notations, we ignore the height of the charging stations, i.e., $h(Q_j) = 0$, $j \in \mathcal{M}$.[1] As mentioned above, the physical movements, including hovering at a specific position to serve users and flying between two positions (serving position and charging position), consume much more energy than data transmission to users [25,26]. Thus, we neglect the energy consumption on transmission and only consider that on physical movements. Let $p_s(z_i)$ and $p_f(v_i)$[2] be the energy consumption per unit time for hovering (serving) at the altitude z_i and flying with speed v_i, respectively. Let t_s be the required time for hovering (serving) for all the UAV-BSs and let t_i be the time of flying of UAV i. Let E be the total energy available at a battery. Let $0 < \alpha < 100\%$ be a given constant representing the percentage of battery that can be used. Then, we have the following constraint:

$$p_s(z_i)t_s + p_f(v_i)t_i \leq \alpha E, \ \forall i \in \mathcal{N}. \tag{9.3}$$

From Eq. (9.3) we can see that the hovering (serving) time is influenced by the flying time. In other words, the longer the distance the UAV-BS has to fly, the less time there will be for hovering (serving). The 3D path for UAV-BS i to

---

[1] It is worth pointing out that if $|\mathcal{M}| \geq |\mathcal{N}|$, we can assume that there is a charging station right under each UAV-BS; otherwise, the positions of charging stations can also be optimized simultaneously with those of UAV-BSs. Obviously, this will make the considered problem even more complex. Thus, we leave this case for further study and only consider the situation with the charging stations at some fixed positions.

[2] Note that the power for flying is influenced by not only the speed, but also the altitude and acceleration. Since the focus of this chapter is not on the energy consumption of UAV-BSs, we adopt a simple model, i.e., the power for flying only depends on the speed.

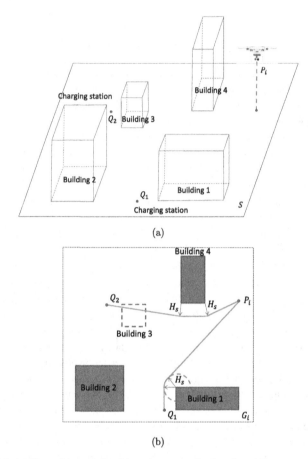

(a)

(b)

FIGURE 9.1 An illustrative example of the trajectory on the virtual map G_i.

fly to charging station j can be divided into two parts. The first part is on G_i, which is the shortest path from the projection of P_i to that of Q_j. We define $g(P_i, Q_j)$ as the length of that shortest path on G_i (see Fig. 9.1 for an example). As depicted in Fig. 9.1a, we have four buildings in the field and the height of Building 3 is more than H_s lower than the altitude of the UAV-BS at P_i. Then, we can obtain a virtual map G_i as shown in Fig. 9.1b. We have two charging stations in the field and the trajectories towards them from the position of UAV are shown in Fig. 9.1b. It is worth mentioning that the trajectories should keep H_s away from buildings. The shortest path with regard to H_s can be computed mathematically by tangent lines. More information on how to find such shortest path can be found in [29] and [30]. The second part is perpendicular to S as well as G_i and the projection of it on S is Q_j, which is given by z_i. To guarantee the hovering (serving) time t_s, we combine these models with Eq. (9.3), and then we have

$$p_s(z_i)t_s + 2p_f(v_i)\frac{g(P_i, Q_j) + z_i}{v_i} \leq \alpha E, \ \forall i \in \mathcal{N}. \tag{9.4}$$

Here, we assume that the speeds for horizontal movement and vertical movement are identical.

To maintain a seamless coverage of the field, we need to have two groups of n UAV-BSs: the first group is serving while the second group is charging if the charging rate is larger than the consuming rate.

We assume that only the outdoor users need to be served by UAV-BSs, while indoor users can be served by indoor SBSs. We assume that a user is considered to be outdoor if the building height at the user's position is zero. An outdoor user at a particular location can have LoS and NLoS with UAV-BSs. We adopt the realistic 3GPP propagation model [26] and have the PL from a UAV-BS to the user as follows:

$$L(P_i, p) = \begin{cases} A^{LoS} + B^{LoS}\log(d(P_i, p)), \\ \text{if the user has LoS link to the UAV,} \\ A^{NLoS} + B^{NLoS}\log(d(P_i, p)), \\ \text{if the user has NLoS link to the UAV,} \end{cases} \tag{9.5}$$

where A^{LoS}, B^{LoS}, A^{NLoS}, and B^{NLoS} are given constants. More information can be found in [26].

We say a user at $p \in S$ is covered by a UAV at P_i if the SINR is above a given threshold, i.e.,

$$SINR(p) = \frac{r_i(PT - L(P_i, p))}{\sum_{j \in \{\mathcal{N}\setminus i\}} r_j(PT - L(P_j, p)) + N_0} \geq \beta, \tag{9.6}$$

where PT is the transmission power of UAV-BSs (which is identical for all UAV-BSs), $PT - L(P_i, p)$ is the power of the received signal at the user from the serving UAV-BS i, the notation r_i describes the Rayleigh fading effect for UAV-BS i and it follows the distribution $r_i \sim \exp(1)$, $\mathcal{N} \setminus i$ is the set of UAV-BSs which are not serving the user, i.e., the set of nonserving UAV-BSs, $\sum_{j \in \{\mathcal{N}\setminus i\}} r_j(PT - L(P_j, p))$ is the total power of the received signal from the nonserving UAV-BSs, which is also called the interference, N_0 is the power of noise, and β is a given constant. Note that the Doppler effect is not considered in this chapter.

Having a high level of SINR at a user does not mean that the user will receive a good user experience. Another influencing factor is the bandwidth allocated to it by the serving UAV-BS. So, we have to consider the limitation of UAV-BSs in terms of capacity. Here we introduce a constant N describing the maximum number of users a UAV can serve. Then, we have the following constraint:

$$\sum_{p \in C(P_i)} \rho(p) \leq N, \ \forall i \in \mathcal{N}, \tag{9.7}$$

where $\rho(p)$ is the user density at position p. In practice, such density can be obtained from statistical data. Here $C(P_i)$ is the set of points of S covered by UAV-BS at P_i.

Let $C(P_1, P_2, \ldots, P_n)$ be the set of points of S covered by the UAV-BSs deployed at $P_1, P_2, \ldots, P_n$, i.e., $C(P_1, P_2, \ldots, P_n) = \cup_{i \in \mathcal{N}} C(P_i)$. In other words, for any $p \in C(P_1, P_2, \ldots, P_n)$, Eq. (9.6) is met for some UAV i. Our objective function, which reflects the total number of covered users by the UAV-BSs, is formulated by

$$f(P_1, P_2, \ldots, P_n) = \sum_{p \in C(P_1, P_2, \ldots, P_n)} \rho(p). \tag{9.8}$$

Therefore, given the space model V, the user density function $\rho(p)$, the height of buildings $h(p)$, the stationary charging stations' locations $Q_1, Q_2, \ldots, Q_{|\mathcal{M}|}$, the power consumed for hovering $p_s()$, the power consumed for flying $p_f(v)$, and constants $H_{min}, H_{max}, H_s, A^{LoS}, B^{LoS}, A^{NLoS}, B^{NLoS}, PT, \alpha, \beta, E, t_s$, and v, we consider the following problem:

$$\max_{P_1, P_2, \ldots, P_n} \sum_{p \in C(P_1, P_2, \ldots, P_n)} \rho(p) \tag{9.9}$$

s.t.

$$h(X, Y) \leq Z, \ \forall \, (X, Y, Z) \in U_i, \forall i \in \mathcal{N}, \tag{9.9a}$$

$$U_i = \{p | d(P_i, p) < H_s, p \in V\}, \forall i \in \mathcal{N}, \tag{9.9b}$$

$$p_s(z_i)t_s + 2p_f(v_i)\frac{g(P_i, Q_j) + z_i}{v_i} \leq \alpha E, \forall i \in \mathcal{N}, \tag{9.9c}$$

$$\sum_{p \in C(P_i)} \rho(p) \leq N, \forall i \in \mathcal{N}, \tag{9.9d}$$

$$x_i \in [X_{min}, X_{max}], \forall i \in \mathcal{N}, \tag{9.9e}$$

$$y_i \in [Y_{min}, Y_{max}], \forall i \in \mathcal{N}, \tag{9.9f}$$

$$z_i \in [H_{min}, H_{max}], \forall i \in \mathcal{N}. \tag{9.9g}$$

9.2.1 Problem difficulty analysis

We analyze the difficulty of the problem of Eq. (9.9) by relaxing some constraints and reducing it to a well-known problem.

- Assume that there are no buildings in the considered area. In this case, every user can have a LoS to a UAV-BS. Then, there is no need to determine whether the UAV-user link is LoS or NLoS.

- Assume that all the UAV-BSs are hovering (serving) at the same height. In this case, constraint (9.9g) is removed and the studied problem turns into a 2D problem.
- Assume that the UAV-BSs use different frequencies to transmit data to users. In this case, there is no interference on users from different UAV-BSs. Then, the SINR is reduced to the SNR. For a given β, we can obtain a certain air-to-ground distance within which the SNR is larger than β. Combining this distance with the height of UAV-BSs, we can compute a coverage radius, and the users which are within such coverage radius to the projection of a UAV-BS can be served by the UAV-BS.
- Assume that the UAV-BSs have sufficient preloaded battery for the operation in the required period. Thus, constraint (9.9c) is removed.
- Assume that the UAV-BSs have sufficient capacity. In this case, constraint (9.9d) is removed.

With the above relaxations, we transform the original optimization problem (9.9) to a problem which can be stated as follows. Given the ground map S, the user density $\rho(p)$, and the coverage radius, find the 2D positions for n UAV-BSs that maximize the total number of covered users. It is clear that this is the well-known k-Max coverage problem, which is NP-hard [31]. Therefore, we can see that the original problem (9.9), without the simplifications, is much more complex than the k-Max coverage problem.

The problem (9.9) is a 3D problem. One challenge is the determination of whether a user at a certain place on S is covered by a UAV-BS. It depends on several factors as shown in Eq. (9.6): the relative distance between the user and the UAV-BS; the local environment, which decides whether they have LoS links; the relative distances between the user and other UAV-BSs; and whether the user has LoS links with other UAV-BSs. So, this 3D problem cannot be addressed by placing circles with a fixed radius on a 2D plane. Also, the positions of UAV-BSs have a mutual impact. Any change of the position of a UAV-BS may result in a different set of covered users of another UAV-BS, due to the change in interference.

9.3 Proposed solution

This section presents a greedy algorithm to address problem (9.9) and discusses its computational complexity.

9.3.1 The greedy algorithm

Now we are in a position to present our solution. We first discretize the continuous space model V into a discrete one by introducing a resolution λ for the horizontal dimension and λ_z for the vertical dimension. Without introducing a new notation, we still use V to represent the space model. Firstly, V is separated into a finite number ($\lceil \frac{H_{max}-H_{min}}{\lambda_z} \rceil$) of planes parallel to the ground and the gap is λ_z. Each plane, such as the ground map S, is further separated into square grids and each edge length is λ. The total number of points on such a plane is $\lceil \frac{X_{max}-X_{min}}{\lambda} \rceil \times \lceil \frac{Y_{max}-Y_{min}}{\lambda} \rceil$. In this discrete space model, the position of a user can be on a point $p \in S$ or not. For simplicity, for each user with a certain pair of coordinates, we use the coordinates of the nearest point $p \in S$ to the user for approximation. If one user has the same distance to more than one such $p \in S$, the corresponding coordinates can be approximated by any of them. Then, the aforementioned user density $\rho(p)$ of the point p can be built up by counting the number of the users which use p's coordinates for approximation. So far, V becomes a set consisting of a finite number of points.

Next, we remove some points from V to obtain a candidate set of points for UAV-BSs. Our removal criteria include the constraints (9.2), (9.4), and (9.9g), because the points which fail to satisfy these constraints are not feasible for UAV-BSs. In particular, the removal according to Eq. (9.9g) is straightforward, i.e., we can simply remove all the points whose heights are under H_{min} from V. The removal according to Eq. (9.2) is also easy for implementation. We can extend the buildings outward by a distance H_s. Then, we remove all the points inside the extended buildings. The removal according to Eq. (9.4) is complex compared to the previous two, because it involves some variables of UAV-BSs, such as the speed v_i. For simplicity, from now on, we assume that all the UAV-BSs fly with the same speed v. Thus, given $p_s()$, v, $p_f(v)$, α, E, t_s, and $Q_1, Q_2, \ldots, Q_{|\mathcal{M}|}$, we can further check a point in V to see whether Eq. (9.4) is satisfied. If not, that point can be removed. Via these removal procedures, we obtain the candidate position set for UAV-BSs and we denote it by $\mathcal{V}$ ($\mathcal{V} \subset V$).

We will also need the following notations to present our algorithm. Let $\mathcal{P}$ be the set of positions of the already selected UAV-BSs, let $\mathcal{C}$ be the set of the already covered points on S by these UAV-BSs, let $\overline{\mathcal{C}}$ be the updated set of covered users by removing those users whose SINRs are below β due to the newly added UAV-BS, let $C(p)$ store the covered users, and let $f(p)$ store the number of totally covered users if the UAV-BS at p is added. Our deployment algorithm is presented in Algorithm 1. The basic idea of this algorithm is to find the position of a UAV-BS in each round from the candidate set $\mathcal{V}$ which contributes the most to the number of the covered users but the number of users covered by each selected UAV-BSs should not exceed the upper bound N, as shown in constraint (9.7). Once n positions have been found or the candidate set $\mathcal{V}$ becomes empty, the algorithm terminates.

Note that when we select the first UAV-BS, we construct the set of covered users based on SNR (line 7), because the positions of the other $n-1$ UAV-BSs are unknown. Correspondingly, the removal procedure in line 8 is not executed for the first UAV-BS. From the second UAV-BS, we recompute the SINRs for the already covered users because adding a new UAV-BS brings extra interference. Some users have to be removed from the covered set because of the decreasing SINR. We then update the set of covered users by each UAV-BS and store the set as $C(\mathcal{P}_k, p)$, if the to be added UAV-BS is at point p. We fur-

---

Algorithm 1 The proposed greedy algorithm.

Require: $\mathcal{V}, \rho(p), N, n$.
Ensure: $\mathcal{P}$.
  1: $\mathcal{P} \leftarrow \emptyset, \mathcal{C} \leftarrow \emptyset, i \leftarrow 1$.
  2: **while** $i \leq n$ **do**
  3:     **if** $\mathcal{V}$ is empty **then**
  4:        Terminate.
  5:     **end if**
  6:     **for** $p \in \mathcal{V}$ **do**
  7:        Construct the set of covered users by the UAV-BS at p, i.e., $C(p)$.
  8:        Add extra interference from the UAV-BS at p to the users in $\mathcal{C}$. Remove the users whose updated SINRs are below β from $\mathcal{C}$ and denote the updated covered set as $\overline{\mathcal{C}}$. Update the set of covered users by each UAV-BS and denote it as $C(P_k, p), \forall k \leq i$.
  9:        **if** $|C(P_k, p)| \leq N, \forall k \leq i$ **then**
10:            $\mathcal{C}(p) \leftarrow \overline{\mathcal{C}} \cup C(p)$.
11:        **else**
12:            $\mathcal{C}(p) \leftarrow \emptyset$.
13:        **end if**
14:        $f(p) \leftarrow \sum_{q \in \mathcal{C}(p)} \rho(q)$.
15:     **end for**
16:     Find p such that $f(p)$ is maximized.
17:     $\mathcal{P} \leftarrow \mathcal{P} \cup p; \mathcal{C} \leftarrow \mathcal{C}(p); C(P_k) \leftarrow C(P_k, p), \forall k \leq i; \mathcal{V} \leftarrow \mathcal{V} \setminus p; i \leftarrow i + 1$.
18: **end while**

---

ther check whether the number of users covered by each UAV-BS exceeds N on line 9. Finally, we compute the number of totally covered users if the to be added UAV-BS is located at p. For all the candidate point in set $\mathcal{V}$, we repeat the above procedures to find the best position which maximizes the total number of covered users.

One significant procedure of Algorithm 1 is the determination of LoS or NLoS between a user and a UAV-BS in lines 7 and 8. Our strategy is to first find a subset of points on S which are (1) within λ to the line segment connecting the user and the projection of a UAV-BS on S and (2) associated with nonzero values in building height. Note that this subset of points is used to approximate the part of the building that is between the user and the UAV-BS. Precisely to check whether the user has the LoS link with the UAV-BS one only needs to check whether there is some part of the building between them. The reason for considering more points within the range of λ is to ensure that we can find a nonempty subset. As λ is the resolution of the grid, this subset should always be nonempty. If we consider the strict case to find points on the line segment connecting the user and the projection of a UAV-BS on S, the subset may be

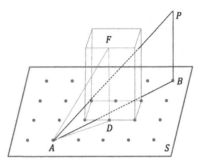

FIGURE 9.2 An illustration for determining LoS or NLoS between a user and a UAV-BS.

empty. Then, we compute the two types of angles. The first type is between the line segment, which connects the user position and the UAV-BS position and S, and the second type is between the line segments, which connect the user position and the positions of the roof points of the building and S. Finally, we compare these two types of angles. Note that there may be more than one point in the selected subset of points for the second type. We say the user has LoS link with the UAV-BS if the former is larger than all the latter angles (or the former is larger than the largest angle among the latter); otherwise, we say the user has NLoS link with the UAV-BS. We present an example in Fig. 9.2, where a user is at point A and a UAV-BS is at point P. The projection of the UAV-BS on S is point B. D is a point on S which is within λ to the line segment AB. Point F is the roof point of the building. Since the angle $\angle PAB$ is smaller than $\angle FAD$, we say the user has NLoS link to the UAV-BS. In other words, the building blocks the LoS between the user and the UAV-BS.

It is worth mentioning that the presented greedy algorithm can obtain suboptimal positions for UAV-BSs. Finding the optimal positions requires locating the n UAV-BSs simultaneously, instead of one by one in the greedy algorithm, which will be much more complex.

9.3.2 Complexity analysis of the greedy algorithm

Finally, we analyze the time complexity of our method. The time complexity of our method depends on the resolutions λ and λ_z. As we consider a 3D scenario, the total number of points in V is proportional to $\frac{1}{\lambda^2 \lambda_z}$. In terms of the removal procedures, the complexity is $O(\frac{1}{\lambda^2 \lambda_z})$, because the removal according to Eq. (9.4) requires to check each point in V once. As we mentioned, not all these points are feasible for UAV-BSs considering the constraints (9.2), (9.4), and (9.9g). The number of points in the candidate set $\mathcal{V}$ depends on the sizes and heights of the buildings in the considered space and the locations of charging stations. It is easy to understand that the larger and taller the buildings, the fewer the candidate positions. Here, we consider an extreme case, i.e., $V = \mathcal{V}$.

To find the position for a UAV-BS in lines 6–15, we need to check each of these $O(\frac{1}{\lambda^2\lambda_z})$ candidates once. On line 7, to construct the cover set for a UAV-BS located at p, we need to check each point on S. The number of points in the set of $C \cup C(p)$ is up to $O(\frac{1}{\lambda^2})$. Further, on line 8 we need to update the covered set by the already selected UAV-BSs and the candidate UAV-BS. Thus, the complexity of lines 6–15 is $O(\frac{n}{\lambda^4\lambda_z})$. Since we need to compute n positions, the overall complexity is $O(\frac{n^2}{\lambda^4\lambda_z})$.

Note that such complexity is for the worst case, and in practice, the computing demand would be much lower. Specifically, the number of candidates in set $\mathcal{V}$ would be much smaller than $O(\frac{1}{\lambda^2\lambda_z})$, due to the existence of buildings as well as the constraints on the positions of UAV-BSs, i.e., Eqs. (9.2), (9.4), and (9.9g). Moreover, the number of points where the users can exist would be much smaller than $O(\frac{1}{\lambda^2})$ and the reason lies in the existence of buildings. However, since both of these two aspects depend on the particular layout of the field map, we cannot provide the theoretical analysis without further assumptions on the shapes, heights, and distribution of buildings in the field, which is beyond the scope of this chapter.

9.4 Performance evaluation

In this section, we evaluate our proposed UAV-BS deployment strategies through simulations. We first set up the simulation environment and then present the simulation results.

9.4.1 Simulation setup

We create an urban area, whose size is 1000 m $\times$ 1000 m, as shown in Fig. 9.3a. The buildings in this field are all cuboids. All the buildings have a width of 60 m and the heights are between 10 and 60 m. This range corresponds to 3- to 20-story buildings, which are commonly seen in urban areas. The street width is 20 m. The altitude limitation for UAV-BSs is between 50 and 300 m. We randomly generate the user density (the number of users) between 0 and 5 on the grid. Furthermore, we manually adjust the user density at certain places to create denser and less dense areas (Fig. 9.3b). Note that the user densities inside buildings are zero. The total number of users in the area is about 3600. All the other parameters used in the simulations are summarized in Table 9.1. In this section, we assume that p_s and p_f are constants and the values are taken from the experimental results presented in [25].

9.4.2 Compared scheme

Although there are some existing publications considering the 3D case such as [19–21], the references [19,20] focus on the deployment of a single UAV-BS while the reference [21] focuses on finding the minimum number of UAV-BSs,

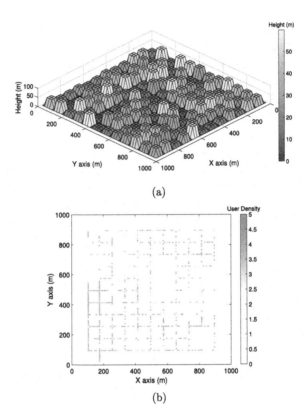

FIGURE 9.3 The simulated urban environment.

TABLE 9.1 Parameter configuration.

Notation	Value	Notation	Value
A^{LoS}	103.8 [26]	B^{LoS}	20.9 [26]
A^{NLoS}	145.4 [26]	B^{NLoS}	37.5 [26]
N_0	-91 dBm [26]	P_t	24, 30 dBm
α	98%	β	15, 20 dB
E	89.2 Wh[a]	v	6 m/s [25]
t_s	0.5 h	H_s	10 m
p_s	140 W [25]	N	500
p_f	143 W [25]		
[a] *https://goo.gl/AKNgq8.*			

which are quite different from this paper. To access the performance gain of the proposed method, we compare it with a scheme proposed in [15]. This scheme focuses on a 2D scenario, i.e., the UAV-BSs are deployed at the same altitude. In this chapter, the altitude is set as 150 m. The objective of [15] is also to deploy a given number of UAV-BSs. Another similarity is that both [15] and this chapter are based on a street graph.

9.4.3 Metrics

We consider two metrics for evaluation:

- Coverage ratio (CR): the ratio of the number of covered users to the total number of users in the field, i.e.,

$$CR = \frac{\sum_{p \in C(P_1, P_2, \dots, P_n)} \rho(p)}{\sum_{p \in S} \rho(p)}. \tag{9.10}$$

- Spectral efficiency (SE): a measure of how efficiently a limited frequency spectrum is utilized by a physical layer protocol. The SE at a user located at $p \in S$ is computed by

$$SE(p) = log_2(1 + SINR(p)). \tag{9.11}$$

Note that in the below evaluations, the SINR of a user is computed based on its actual location.

We use the metric of average SE (ASE):

$$ASE = \frac{\sum_{p \in C(P_1, P_2, \dots, P_n)} \rho(p) SE(p)}{\sum_{p \in C(P_1, P_2, \dots, P_n)} \rho(p)}. \tag{9.12}$$

Note that the existing LTE network with Turbo codes can achieve a performance very close to the Shannon capacity law [32].

9.4.4 Simulation results

In the created environment, we conduct various simulations to study the performance of our proposed approach.

9.4.4.1 Algorithm evaluation

We first investigate the algorithm performance against the resolution λ and λ_z. The number of UAV-BSs is fixed at 10, λ is from 10 to 100, and λ_z takes values of 10, 20, and 30. $PT = 30$ dBm and $\beta = 15$ dB. The results are shown in Fig. 9.4a. It is easy to understand that with the increase of λ and λ_z, the CR decreases. We study the gap between the greedy algorithm and the exact solution by brute-force search (exhaustive search). We set $\lambda = 50$ and $\lambda_z = 30$ and the number of UAV-BSs ranges from 1 to 5. For a larger number of UAV-BSs, the brute-force search cannot obtain the exact solution in a reasonable time. The comparison is shown in Fig. 9.4b. It is clear that for a small number of

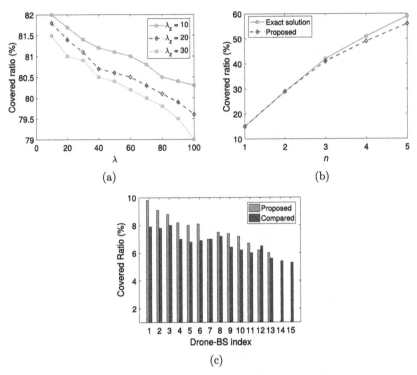

(a)

(b)

(c)

FIGURE 9.4 (a) Covered ratio versus the resolution. (b) Comparison with the exact solution. (c) The covered ratios by the proposed method and the method reported in [15].

UAV-BSs, the proposed method achieves a close CR with the exact solution. Furthermore, we compare with [15] to assess the performance of the proposed method. In the simulated environment, we keep adding new UAV-BSs according to Algorithm 1, until no more UAV-BSs need to be added to cover users. For each number of UAV-BSs in the compared method, we conduct 10 simulations with different initial positions of UAV-BSs and take the best result among them. The comparison is shown in Fig. 9.4c. As the UAV-BSs are at the same altitude, the performance of [15] is worse than the proposed method, and the number of UAV-BSs to cover all the users in [15] is also larger than the proposed one. Furthermore, the CRs of each UAV-BS in the two methods are shown in Fig. 9.4c. In general, the UAV-BSs in the proposed method cover more users than the compared one. For the proposed method, the maximum percentage of users served by a UAV-BS is 9.8% while that of the compared approach is 8%; the minimum percentage of the proposed method is 6% while that of the compared approach is 5.3%; and the average percentage of the proposed method is 7.7% while that of the compared approach is 6.7%.

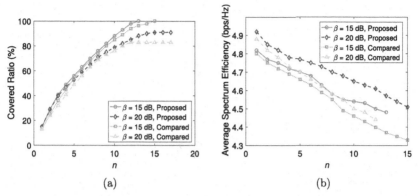

FIGURE 9.5 The influence of the SINR threshold on coverage ratio and average spectral efficiency.

9.4.4.2 Parameter impacts

Now, we consider the influence of some key parameters. We first study the impact of the SINR threshold β. We keep the transmission power PT as 30 dBm. Fig. 9.5 shows the CR and the ASE, with different numbers of UAV-BSs (n) and different values for β. In general, with the increase of n, the CR increases, while the ASE decreases. It is easy to understand that when there are more UAV-BSs in the field, more users can be covered. At the same time, the interference effect is enhanced. Thus, the SINRs at users reduce for the given transmission power, which leads to lower ASE according to Eqs. (9.11) and (9.12). We can also see from Fig. 9.5a that for $\beta = 20$ dB, it is impossible to cover all the users in the field. However, when β is 15 dB, 13 and 15 UAV-BSs are needed to cover all the users in the simulated environment by the proposed scheme and the compared one, respectively. Furthermore, for the same number of UAV-BSs, increasing β means reducing the cover range in general, which leads to a decrease in CR and an increase in ASE. Fig. 9.6 compares the results of the CR and the ASE with different transmission powers (PT), where the SINR threshold is set as $\beta = 15$ dB. PT has a significant impact on the system, because it directly influences the SINR, and then indirectly influences the CR. From Fig. 9.6 we can see that for a given number of UAV-BSs (n), in general, larger PT leads to a higher CR and ASE. The increase of PT increases the received power as well as the interference. From the numerical results, we can see that the increase of received power at users is larger than that of the interference, thus SINRs are increased and more users can be covered, and the ASE is increased.

In summary, from the numerical results, we can see that the proposed greedy algorithm achieves similar performance with the exact solution and outperforms the compared scheme [15] proposed by the same authors, which considers a 2D case. Moreover, a higher SINR threshold leads to higher SE but reduces the CR for a certain number of UAV-BSs. Higher transmission power is promising to increase both the CR and the SE, but it consumes more energy. These factors

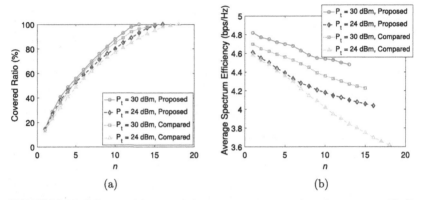

FIGURE 9.6 The influence of the transmission power on coverage ratio and average spectral efficiency.

need to be accounted for comprehensively in designing a cellular network making use of UAV-BSs.

9.5 Summary

In this chapter, we considered the scenario of using autonomous UAV-BSs to serve the ground outdoor users in cellular networks. We studied the problem of 3D deployment of UAV-BSs in a given area to maximize the number of covered users and in the meanwhile, each covered user achieves an acceptable user experience, with the consideration of the safety issue of UAV-BSs. We formulated an optimization problem, analyzed the difficulty of such a problem, and presented a greedy algorithm together with the time complexity analysis. On a created urban environment with a certain distribution of buildings, we conducted extensive simulations to investigate the influence of various parameters, including the number of UAV-BSs, the SINR threshold, the transmission power level, and the heights of buildings, on the performance of the proposed approach, in terms of CR and ASE. Finally, a comparison with a baseline algorithm is provided to assess the performance gains of the proposed approach.

One limitation of the presented work is that the traffic demand is assumed to be known in advance, while in practice the users are mobile which makes the service demand quite dynamic. Therefore, one future research direction is to release such an assumption and find the positions for UAV-BSs to provide service to mobile users. Another shortcoming is that environmental factors such as rain or wind have not been considered in this chapter. These issues need further study.

References

[1] H. Huang, A.V. Savkin, A method for optimized deployment of unmanned aerial vehicles for

maximum coverage and minimum interference in cellular networks, IEEE Transactions on Industrial Informatics 15 (5) (2019) 2638–2647.

[2] V. Sharma, K. Srinivasan, H.-C. Chao, K.-L. Hua, W.-H. Cheng, Intelligent deployment of UAVs in 5G heterogeneous communication environment for improved coverage, Journal of Network and Computer Applications 85 (2017) 94–105.

[3] A. Al-Hourani, S. Kandeepan, A. Jamalipour, Modeling air-to-ground path loss for low altitude platforms in urban environments, in: IEEE Global Communications Conference (GLOBECOM), 2014, pp. 2898–2904.

[4] A. Al-Hourani, S. Kandeepan, S. Lardner, Optimal LAP altitude for maximum coverage, IEEE Wireless Communications Letters 3 (Dec 2014) 569–572.

[5] I. Bor-Yaliniz, S.S. Szyszkowicz, H. Yanikomeroglu, Environment-aware drone-base-station placements in modern metropolitans, IEEE Wireless Communications Letters 7 (3) (2018) 372–375.

[6] Propagation data and prediction methods for the planning of short-range outdoor radio communication systems and radio local area networks in the frequency range 300 mHz to 100 GHz, 2017.

[7] D. González, H. Hakula, A. Rasila, J. Hämäläinen, Spatial mappings for planning and optimization of cellular networks, IEEE/ACM Transactions on Networking 26 (Feb 2018) 175–188.

[8] X. Li, D. Guo, H. Yin, G. Wei, Drone-assisted public safety wireless broadband network, in: IEEE Wireless Communications and Networking Conference Workshops (WCNCW), March 2015, pp. 323–328.

[9] S. Rohde, M. Putzke, C. Wietfeld, Ad hoc self-healing of OFDMA networks using UAV-based relays, Ad Hoc Networks 11 (Sep 2013) 1893–1906.

[10] V. Sharma, M. Bennis, R. Kumar, UAV-assisted heterogeneous networks for capacity enhancement, IEEE Communications Letters 20 (June 2016) 1207–1210.

[11] J. Lyu, Y. Zeng, R. Zhang, T.J. Lim, Placement optimization of UAV-mounted mobile base stations, IEEE Communications Letters 21 (March 2017) 604–607.

[12] D. Zorbas, L.D.P. Pugliese, T. Razafindralambo, F. Guerriero, Optimal drone placement and cost-efficient target coverage, Journal of Network and Computer Applications 75 (2016) 16–31.

[13] A.V. Savkin, H. Huang, Asymptotically optimal deployment of drones for surveillance and monitoring, Sensors 19 (9) (2019) 2068.

[14] A.V. Savkin, H. Huang, Proactive deployment of aerial drones for coverage over very uneven terrains: a version of the 3D art gallery problem, Sensors 19 (6) (2019) 1438.

[15] H. Huang, A.V. Savkin, An algorithm of efficient proactive placement of autonomous drones for maximum coverage in cellular networks, IEEE Wireless Communications Letters 7 (6) (2018) 994–997.

[16] A.V. Savkin, H. Huang, A method for optimized deployment of a network of surveillance aerial drones, IEEE Systems Journal 13 (4) (2019) 4474–4477.

[17] B. Galkin, J. Kibilda, L.A. DaSilva, Deployment of UAV-mounted access points according to spatial user locations in two-tier cellular networks, in: Wireless Days (WD), 2016, IEEE, 2016, pp. 1–6.

[18] A.V. Savkin, H. Huang, Deployment of unmanned aerial vehicle base stations for optimal quality of coverage, IEEE Wireless Communications Letters 8 (1) (2019) 321–324.

[19] R.I. Bor-Yaliniz, A. El-Keyi, H. Yanikomeroglu, Efficient 3-D placement of an aerial base station in next generation cellular networks, in: IEEE International Conference on Communications (ICC), 2016, pp. 1–5.

[20] M. Alzenad, A. El-Keyi, F. Lagum, H. Yanikomeroglu, 3D placement of an unmanned aerial vehicle base station (UAV-BS) for energy-efficient maximal coverage, IEEE Wireless Communications Letters 6 (Aug 2017) 434–437.

[21] E. Kalantari, H. Yanikomeroglu, A. Yongacoglu, On the number and 3D placement of drone base stations in wireless cellular networks, in: IEEE Vehicular Technology Conference (VTC-Fall), 2016, pp. 1–6.

[22] H. Huang, A.V. Savkin, An algorithm of reactive collision free 3-D deployment of networked unmanned aerial vehicles for surveillance and monitoring, IEEE Transactions on Industrial Informatics 16 (1) (2020) 132–140.

[23] K.A. Swieringa, C.B. Hanson, J.R. Richardson, J.D. White, Z. Hasan, E. Qian, A. Girard, Autonomous battery swapping system for small-scale helicopters, in: International Conference on Robotics and Automation (ICRA), IEEE, 2010, pp. 3335–3340.

[24] K.A. Suzuki, P. Kemper Filho, J.R. Morrison, Automatic battery replacement system for UAVs: analysis and design, Journal of Intelligent & Robotic Systems 65 (1–4) (2012) 563–586.

[25] A. Fotouhi, M. Ding, M. Hassan, Understanding autonomous drone maneuverability for internet of things applications, in: IEEE International Symposium on a World of Wireless, Mobile and Multimedia Networks (WoWMoM), June 2017, pp. 1–6.

[26] 3GPP TR 36.828: further enhancements to LTE time division duplex (TDD) for (DL-UL) interference management and traffic adaptation, 2012.

[27] T. Bai, R.W. Heath, Coverage and rate analysis for millimeter-wave cellular networks, IEEE Transactions on Wireless Communications 14 (Feb 2015) 1100–1114.

[28] H. Huang, A.V. Savkin, M. Ding, M.A. Kaafar, Optimized deployment of drone base station to improve user experience in cellular networks, Journal of Network and Computer Applications 144 (2019) 49–58.

[29] A.V. Savkin, H. Huang, Optimal aircraft planar navigation in static threat environments, IEEE Transactions on Aerospace and Electronic Systems 53 (Oct 2017) 2413–2426.

[30] H. Huang, A.V. Savkin, Viable path planning for data collection robots in a sensing field with obstacles, Computer Communications 111 (Oct 2017) 84–96.

[31] U. Feige, A threshold of ln n for approximating set cover, Journal of the ACM (JACM) 45 (July 1998) 634–652.

[32] S. Sesia, M. Baker, I. Toufik, LTE-the UMTS Long Term Evolution: From Theory to Practice, John Wiley & Sons, 2011.

Chapter 10

Energy-efficient path planning of solar-powered UAVs for communicating with mobile ground users in urban environments[☆]

10.1 Motivation

Since using a single UAV is often inefficient to conduct a complex mission, employing a UAV team is the trend in order to complete missions quickly [1]. In the past few decades, the coordination issue of multiagent systems has attracted great attention from different research communities [2–4]. This chapter focuses on serving moving users by a group of UAVs. A practical application of the considered scenario is that in WSNs, the sensor nodes collect data from the environment. UAVs function as data sinks to collect the sensory data from sensor nodes [5]. In general, the number of available UAVs is smaller than the number of sensor nodes. Thus, the UAVs carry out periodical surveillance of the sensor nodes.

Since UAVs often have limited onboard battery capacity, their operation duration is constrained. Installing solar panels enables the UAVs to harvest energy from the sun, which is promising to prolong the lifetime of the UAVs in the daytime [6]. We consider the surveillance of mobile users by the solar-powered UAVs in urban environments. The tall buildings have some negative impact on the mission. Firstly, they may create some shadow region where the LoS between the UAVs and the sun is blocked, so that the UAVs cannot harvest energy. Besides, the buildings may also block the LoS between the UAVs and the considered users. This may prevent a UAV from successfully surveying a user.

The problem of interest is formulated as an optimization problem to minimize the user revisit time by planning the UAVs' paths. To address this problem, we present an RRT-based path planning method. This method can quickly find a feasible path for the UAV to intercept the user in the scenario where the user

☆ The main results of the chapter were originally published in Hailong Huang, Andrey V. Savkin, Energy-efficient autonomous navigation of solar-powered UAVs for surveillance of mobile ground users in urban environments, Energies 13 (21) (2020) 5563.

Copyright © 2022 Elsevier Inc. All rights reserved.

moves along a known trajectory. We then consider the case with one UAV and multiple users. We present a nearest neighbor-based navigation method. The so-called nearest neighbor involves both the UAV–user distance and the uncertainty level of a user. Finally, we consider the multi-UAV and multiuser case. We partition the users into groups according to the distance information between the users and the UAVs. Then, each UAV takes care of the users in its partition.

The proposed autonomous navigation algorithm which navigates a UAV team to periodically survey a group of mobile ground users is the main contribution of this chapter. It is computationally efficient and is easily implementable online, since it is a randomized RRT-based approach. Extensive simulation results are reported to confirm the effectiveness of the developed method. The main results of the chapter were originally published in [6].

The remainder of the chapter is organized as follows. Section 10.2 briefly reviews the relevant work. Section 10.3 presents the system models and formulates the problem. Section 10.4 presents the proposed UAV navigation approaches. Section 10.5 reports the computer simulation results, and Section 10.6 gives the concluding remarks.

10.2 Related work

The user surveillance problem considered in this chapter has not been considered in any existing publications. In this section, we present some closely related publications so that we can distinguish the contributions of the chapter from others.

The target surveillance problem has been investigated from different levels in the literature. In terms of sensing, a large number of image/video processing strategies have been developed to estimate the states of the users from the measured images/videos [7–11]. In these publications, attention has been paid to the quality of detection for a single target.

In the scenario with multiple users, UAV resource allocation becomes necessary to achieve a good quality of surveillance. Many operational research results such as the conventional TSP [12] and the VRP [13] are commonly used tools to plan the UAVs' paths. When there are enough UAVs, the coverage control has been investigated to achieve the optimal sensing coverage of the users [14,15]. In cases where moving users are to be monitored, reactive algorithms have been proposed to maintain the quality of sensing [16,17].

This chapter focuses on the scenario where the number of UAVs is not enough to persistently monitor the users, so periodical surveillance is conducted by the UAVs. As the users are moving in the considered region, the problem is more relevant to the time-dependent TSP [18] and the moving-target TSP [19,20]. In the time-dependent TSP [18], the common setting is to find the shortest tour for the salesman in a graph with time-dependent edges. In the moving-target TSP [19,20], the users are assumed to move with a constant speed along a fixed direction. The problem considered here is different from them. Firstly, the users move along some trajectories so that their speeds and moving directions may change with time. Secondly, the existence of buildings in the urban environment requires the UAVs to avoid collision with the buildings. Thirdly, to enable the UAVs to operate in the given time period, the UAVs need to harvest energy from the sun. However, each path depends on the UAV's initial position, the buildings' positions, and the target's trajectory, which is challenging to be known in advance. Thus, both the moving-target TSP and the time-dependent TSP cannot be applied to address the problem considered in the chapter.

Path planning plays an important role in this work. Among various path planning algorithms, RRT is a sampling-based approach. It generates a feasible (but possibly suboptimal) path quickly even if the environment is complicated. Many publications have reported that this method can be easily applied in real-time applications such as mobile ground robots [21] and autonomous driving [22]. To improve the solution quality and computing speed, many RRT variants have been developed. Special attention has been paid to the generation of samples and the control of the searching step length. A lower bound tree-RRT is designed to obtain the near-optimal path [23]. Besides, a node control strategy is proposed to restrict the expansion of the random tree [24]. Due to the computational efficiency, RRT-based approaches are generally suitable to run in real-time, and they also have the potential to be implemented in a decentralized manner [25]. In this chapter, we adopt the RRT approach. However, as will be shown in the following sections, we do not have a fixed destination for a UAV. Instead, the destination of a UAV moves. Our objective is to generate a feasible path in real-time so that the UAV can intercept the target as soon as possible.

10.3   System model and problem statement

Suppose we have a team of solar-powered UAVs labeled $1, 2, \ldots, I$. We consider that these solar-powered UAVs fly at a fixed altitude z in an urban area to conduct some missions. For UAV i, let $p_i(t) = [x_i(t), y_i(t), z]$ be its position in the ground frame at t, let $\theta_i(t)$ be the horizontal heading angle with respect to the x-axis, and let $v_i(t)$ and $\omega_i(t)$ be its linear and angular speeds, respectively. The motion of UAV i can be described by the following equations [26,27]:

$$\begin{cases} \dot{x}_i(t) = v_i(t)\cos(\theta_i(t)), \\ \dot{y}_i(t) = v_i(t)\sin(\theta_i(t)), \\ \dot{\theta}_i(t) = w_i(t). \end{cases} \tag{10.1}$$

In this chapter, the effect of wind has not been considered. The following constraints hold for any UAV at any time:

$$\begin{cases} -V_i^{\max} \le v_i(t) \le V_i^{\max}, \\ -\Omega_i^{\max} \le w_i(t) \le \Omega_i^{\max}, \\ (x_i(t), y_i(t)) \in \mathcal{D}. \end{cases} \tag{10.2}$$

TABLE 10.1 Symbols and meanings.

Symbol	Meaning
$p_i(t)$	UAV i's position at time t
$\theta_i(t)$	UAV i's heading angle
$v_i(t)$	UAV i's linear speed
$\omega_i(t)$	UAV i's angular speed
α	Observation angle
R	Radius of the vision cone
$P_i(t)$	UAV i's consuming power
$P_i^{sun}(t)$	UAV i's harvesting power
$Q_i(t)$	UAV i's residual energy
$q_j(t)$	User j's position
τ_j	User j's revisit time
$d_{ij}(t)$	Horizontal distance between UAV i and user j

Here, Ω_i^{max} and V_i^{max} are the given constants and $\mathcal{D} \subset \mathbf{R}^2$ is the considered area. The movement of many UAVs can be described by (10.1) and (10.2); see [28–31]. It is worth pointing out that in (10.2), the linear speed $v_i(t)$ can take a negative value. This allows a UAV to move backwards when necessary. In Section 10.5, we will see some UAV trajectories with sharp turns, and the reason is that a negative linear speed is applied. This avoids making a large turn by moving along a circle. The frequently used symbols in the chapter are summarized in Table 10.1, together with their meanings.

Let $P_i^{sun}(t)$ be the harvesting power of the solar energy. It can be computed as follows [32]:

$$P_i^{sun}(t) = \eta A_i \cos\phi(t), \tag{10.3}$$

where η represents the solar cell efficiency, A_i represents the area of the solar cells, and ϕ gives the incidence angle. The incidence angle ϕ is further dependent on the azimuth angle α_z and the elevation angle α_e of the sun, and in the daytime, both α_z and α_e vary with time.

The UAVs consume energy when they are flying. For the energy-consuming power, we follow the model used in [33]:

$$P_i(t) = \lambda_0 \left(1 + \frac{3v_i^2}{U_{tip}^2}\right) + \frac{1}{2}\rho\kappa s S v_i^3(t) + \lambda_1 \left(\sqrt{1 + \frac{v_i^4(t)}{4\mu^4}} - \frac{v_i^2(t)}{2\mu^2}\right)^{\frac{1}{2}}, \tag{10.4}$$

where λ_0, λ_1, and μ are the blade profile power, the induced power, and the mean rotor-induced velocity in hovering, respectively, U_{tip} represents the tip speed of the rotor blade, κ is the fuselage drag ratio, s represents the rotor solidity, ρ is the air density, and S_i is the rotor disc area.

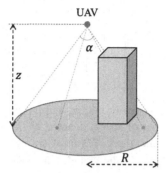

FIGURE 10.1 The observation of the ground facing camera. The user at the green (mid gray in print version) point can be seen by the UAV while the one at the red (light gray in print version) point cannot be seen due to the blockage of the building.

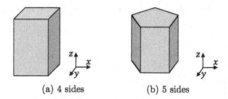

(a) 4 sides   (b) 5 sides

FIGURE 10.2 Prisms.

Let $Q_i(t)$ denote the residual energy of the battery of UAV i. We have

$$\dot{Q}_i(t) = P_i^{sun}(t) - P_i(t). \tag{10.5}$$

Moreover, Q_i^{max} represents the upper bound of $Q_i(t)$.

Each UAV carries a ground facing camera; the camera's observation angle is denoted by $\alpha \in (0, \pi)$ (Fig. 10.1). If a user is located in a disc centered at $p_i(t)$ of radius

$$R = z \tan\left(\frac{\alpha}{2}\right) \tag{10.6}$$

and has LoS with the UAV, it can be observed by the UAV. We assume that a gimbal is available on the UAV, so that no matter how the UAV moves, the camera always faces the ground.

Now, we model the buildings. In this chapter, each building is modeled as the smallest prism to enclose this building. Each prism has two parallel and congruent bases and a number of flat sides which are perpendicular to the xy-plane (Fig. 10.2). Each prism is characterized, Ψ, ψ, and h, where Ψ is a κ-by-2 matrix and ψ is a κ-by-1 vector, which determine the shape and size of the base, and h is a scalar indicating the height of the prism. For a point (x, y, z), if it is inside a prism, we have

$$\begin{cases} \Psi \begin{bmatrix} x \\ y \end{bmatrix} \leq \psi, \\ 0 \leq z \leq h. \end{cases} \tag{10.7}$$

Given the environment information, Ψ, ψ, and h are known for each building. Then, we can have a subset of space, denoted by $\mathcal{X}_{building}$, corresponding to these buildings. At any time, the UAVs must not be inside $\mathcal{X}_{building}$. Clearly, avoiding $\mathcal{X}_{building}$ is similar to the collision avoidance with steady obstacles [34,35].

Having the model of buildings is not sufficient to characterize the observation region of a UAV. We also need a method to determine whether a position in the air and a position on the ground have LoS. For this purpose, we consider the straight-line segment between two points p and q, which is described as

$$\begin{cases} x = x_q + \beta_x \tau, \\ y = y_q + \beta_y \tau, \\ z = z_q + \beta_z \tau, \\ \min\{x_p, x_q\} \leq x_q + \alpha \tau \leq \max\{x_p, x_q\}, \end{cases} \tag{10.8}$$

where $q = (x_q, y_q, z_q)$, $p = (x_p, y_p, z_p)$, $(\beta_x, \beta_y, \beta_z) = \frac{\overrightarrow{pq}}{\|\overrightarrow{pq}\|}$, and τ is a free variable.

Whether p and q have LoS can be tested by looking for the intersection points between the line segment connecting p and q (10.8) and any prism (10.7). As the environment information is known, whether p and q have LoS can be easily confirmed. We introduce a function $LOS(p, q, \mathcal{X}_{building})$:

$$LOS(p, q, \mathcal{X}_{building}) = \begin{cases} 1, & \text{if } p \text{ and } q \text{ have LoS}, \\ 0, & \text{otherwise.} \end{cases} \tag{10.9}$$

With this function, we can also test if a UAV and the sun have LoS. To this end, the sun's location needs to be known. Let V, a unit vector, denote the sunlight direction. With V, we can imagine that the sun is at $q_{sun} = p - V\tau$, where the parameter τ takes a large value so that the sun is far from the point p. We need to place the sun at a relatively far position. The reason is that we use the line segment to verify whether two points have LoS. When the sun is placed close to the point p, we may not obtain the correct verification.

Let $b_i(t)$ be a binary variable indicating whether UAV i has LoS with the sun. Then, UAV i's residual energy can be computed by

$$\dot{Q}_i(t) = P_i^{sun}(t) b_i(t) - P_i(t). \tag{10.10}$$

There are N mobile ground users in the given urban environment to be periodically surveyed. These users can be some sensor nodes to measure the environment information of interest. Instead of continuously transmitting the sensory data, they only upload their sensory data to a control unit via the UAVs in proximity. This setting can prolong the lifetime of the nodes when the sensory data are large, such as videos. We assume that the UAVs know the current positions of the users, and this information can be provided by the users since the energy consumption of reporting the position information can be ignored compared to the large size of sensory data. We also assume that the users' future positions are predictable. This assumption is reasonable since when the users carry out some predefined missions, their trajectories can be known. Let $q_j(t) \in \mathbf{R}^3$ denote user j's location ($j = 1, \ldots, N$) at time t.

In this chapter, we consider that $I < N$ and the users spread in the considered environment. Then, there may be some time in which a user is not under surveillance. Based on common sense, the uncertainty level of a user is related to the time in which it is not under surveillance. Thus, a significant objective of the surveillance system is to maintain the uncertainty level of the users as low as possible. This can be formulated as the minimization of the maximum user revisit time. Let τ_j denote the time gap between two consecutive visits of user j. Let $d_{ij}(t)$ denote the horizontal distance between user j and UAV i at time t.

Definition 10.3.1. User j is under service by UAV i at time t if $LOS(p_i(t), q_j(t), \mathcal{X}_{building}) = 1$ and $d_{ij}(t) \leq R$.

Let $s_j(t)$ be a binary variable indicating if user j is under surveillance at time t. Then, we have

$$s_j(t) = \begin{cases} 1, & \text{if } \exists i \text{ such that } LOS(p_i(t), q_j(t), \mathcal{X}_{building}) = 1 \text{ and } d_{ij}(t) \leq R, \\ 0, & \text{otherwise.} \end{cases}$$

(10.11)

Then, we can use $s_j(t)$ to calculate τ_j. Specifically, we have

$$\tau_j = \max_{s_j(t_1)=1, s_j(t_2)=1, t_1 < t_2, \forall t_1 < t < t_2, s_j(t)=0} t_2 - t_1.$$

(10.12)

In other words, τ_j is the time instant gap between the two consecutive visits. Note that if there is only one visit during the mission period $[0, T]$, i.e., at t_1, then, $\tau_j = T - t_1$. If there is not any visit during the mission period, then $\tau_j = T$.

The problem under investigation is to develop a navigation method for the UAVs modeled by (10.1) and (10.2) to minimize the maximum revisit time during the mission period $[0, T]$, i.e.,

$$\min \max_{j=1,\ldots,N} \tau_j$$

(10.13)

s.t.

$$Q_i(t) > 0, \ \forall i, \ \forall t \in [0, T].$$

(10.14)

The considered problem is difficult to address optimally. Although we can have the trajectories of users and predict their positions for the period of $[0, T]$, it is still hard to plan the trajectories of the UAVs in advance. The main reason lies in the complexity of the flying space in urban environments. In particular, due to the existence of buildings, the trajectory of UAV i (suppose it is assigned to serve user j) depends on both the trajectory of user j and the position of UAV i at which it is assigned with this task. Furthermore, the UAV's position depends on its last task. The coupling of the high-level task assignment problem and the low-level trajectory planning problem makes it too complex to be addressed optimally. Even if an optimal solution can be obtained, it may also take so long that it cannot be applied online.

10.4 Proposed navigation method

10.4.1 Lifetime of UAVs

Before presenting the navigation method, it is necessary to discuss the lifetime of the UAVs. As shown in (10.3), for a UAV, its energy harvesting power varies with time because of the movement of the sun. Then, the maximum harvested energy amount by UAV i is given by $\int_0^T \eta A_i \cos \phi(t) dt$.

According to (10.4), the energy-consuming power P_i increases with the speed v_i in the first and second terms, while it decreases with v_i in the third term. Additionally, the third term weighs more when v_i is relatively small, while the first and second terms weigh more when v_i is relatively large. Thus, the energy-consuming power P_i in (10.4) firstly decreases and then increases with v_i. An illustrative plot of the relationship between P_i and v_i is shown in Fig. 10.3. There exists an optimal linear speed such that the energy consuming power is minimized. Let P_i^{opt} denote the minimum energy-consuming power.

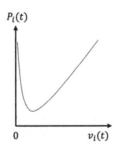

FIGURE 10.3 Illustration of the relationship between $P_i(t)$ and $v_i(t)$.

We assume that Q_i^{max} is the initial energy of UAV i. The necessary condition for the UAV to conduct the surveillance mission is as follows:

$$Q_i^{max} + \int_0^T \eta A_i \cos \phi(t) dt - T P_i^{opt} > 0. \qquad (10.15)$$

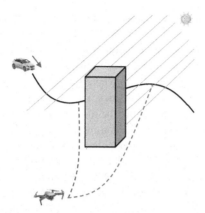

FIGURE 10.4 Illustration of the potential benefit offered by the margin energy capacity. The margin energy capacity allows the UAV to take a path with some part having no LoS with the sun to reduce the revisit time.

Once (10.15) holds, the UAV can complete the surveillance mission with the optimal linear speed, though this does not guarantee the performance of surveillance. Moreover, a hidden assumption of (10.15) is that the UAV always has LoS with the sun in $[0, T]$. Otherwise, the harvested energy amount is smaller than $\int_0^T \eta A_i \cos\phi(t)dt$. However, if (10.15) does not hold, the UAV cannot conduct the mission. If the capacity Q_i^{max} takes some larger value, it is allowed to have some part of the UAV trajectory having no LoS with the sun. The margin of capacity has the potential to reduce the revisit time. As shown in Fig. 10.4, there are two feasible paths for the UAV to intercept the user. The red (dark gray in print version) one has some part behind the building that prevents the UAV from harvesting energy, while the green (mid gray in print version) path enables the UAV to harvest energy during the trip. The red one leads to a shorter time for the UAV to serve the user than the green one. If there is no margin capacity, the UAV has to follow the green path. Otherwise, it can follow the red one to reduce the revisit time.

Though the margin energy capacity brings benefits in terms of the reduction of the revisit time, it also creates challenges in the management of this amount of energy. Specifically, the UAV may need to survey several users during the mission. Then, it is difficult to decide how to allocate the margin capacity to the tasks. In the subsequent parts, we assume that there is no such margin capacity, and the UAVs fly at their corresponding optimal linear speeds. So, the UAVs always look for paths having LoS with the sun. We leave the complex case with margin capacity for future study.

10.4.2 One UAV and one user

In the simplest case, we consider that one UAV is assigned to survey a user. Suppose that at t_0 the UAV starts to move to the user. We denote the initial positions of the UAV and the user as $p(t_0)$ and $q(t_0)$, respectively, and the initial heading angle of the UAV as $\theta(t_0)$. We use $\mathbf{x}(t) = [p(t), \theta(t)]$ to represent the state of the UAV. Suppose we know the trajectory of the user in the considered area $\mathcal{D}$, and we make a prediction for its future positions in the time interval $[t_0, t_0 + T_0]$. Formally, we know $q(t)$ where $t \in [t_0, t_0 + T_0]$. We select a suitable T_0 so that the UAV can intercept the user before $t_0 + T_0$.

For the purpose of implementing the method online, we adopt the computationally efficient RRT approach. The common setting of the RRT approach looks for a feasible path between a start position and a destination, and the obtained path can avoid obstacles. In our problem, the buildings taller than the UAV's flying altitude are regarded as obstacles. Our problem has some additional features compared to the common setting. Firstly, different from a stationary destination, the destination in our problem, i.e., the user, is moving. Secondly, the UAV should avoid shadow regions, because it needs to always harvest energy during the trip.

The objective is to find a feasible UAV path such that at time instant $t \in [t_0, t_0 + T_0]$, the user is located inside the vision cone of the UAV and they have LoS. Starting from the current UAV position, i.e., $p(t_0)$, we generate a random tree. The termination condition of the tree generation process is that $d(t) \leq R$ and $LOS(p(t), q(t), \mathcal{X}_{building}) = 1$ at time t, where $d(t)$ is the horizontal UAV–user distance.

We present all the procedures in Algorithm 1. Let $\mathcal{T}$ denote the random tree and let δ be a sampling interval; $\mathcal{T}$ consists of a set of vertices. A vertex is annotated with control inputs, parent vertex, and timestamp. Firstly, we initialize the tree $\mathcal{T}$ with $\mathbf{x}(t_0)$. Then, we keep generating random samples in the space, find the nearest vertex from the tree to the sample, and choose the suitable control inputs to generate a new vertex $\mathbf{x}_{new}$. We further check whether $\mathbf{x}_{new}$ belongs to

---

Algorithm 1 RRT-based path planning algorithm to intercept a user.

Input: $\mathbf{x}(t_0)$, $\mathcal{X}_{building}$, $q(t)$ where $t \in [t_0, t_0 + T_0]$
Output: $\mathcal{T}$
Initialize $\mathcal{T}$
while There do not exist a timestamp k and a vertex $\mathbf{x}^*$ with the timestamp of k, such that $d(t_0 + k\delta) \leq R$ and $\mathbf{x}^*$ has LoS with $q(t_0 + k\delta)$ **do**
    Randomly generate a node in the space.
    Select the closest vertex from $\mathcal{T}$ to the node.
    Choose the appropriate control inputs to generate $\mathbf{x}_{new}$, such that $\mathbf{x}_{new}$ cannot be closer to the node after δ.
    if $\mathbf{x}_{new} \notin \mathcal{X}_{building}$ and has LoS with the sun **then**
        Associate with $\mathbf{x}_{new}$ its parent, the applied control inputs and timestamp, and add $\mathbf{x}_{new}$ to $\mathcal{T}$.
    end if
end while

$\mathcal{X}_{building}$ and whether it has LoS with the sun. When both conditions are verified, we associate the parent vertex, the control inputs, and the timestamp with this vertex, and add it to the random tree. Here, the timestamp is an integer indicating the number of steps from the initial vertex to this vertex. We stop growing the tree once there exist a timestamp k and a vertex $\mathbf{x}^*$ with the timestamp of k, such that $d(t_0 + k\delta) \leq R$ and the position of $\mathbf{x}^*$ has LoS with $q(t_0 + k\delta)$.

When the tree growth process terminates, we obtain the final vertex on the UAV path, i.e., $\mathbf{x}^*$. We also know that it takes k steps for the UAV from its initial position to reach $\mathbf{x}^*$. By a standard backtracking algorithm, we can find the path from $\mathbf{x}^*$ back to the initial position. An example is shown in Fig. 10.5, where t_0 is set as 0. In this example, we stop growing the random tree since at time 3δ, the condition $d(3\delta) \leq R$ holds (suppose $LOS(p(3\delta), q(3\delta), \mathcal{X}_{building}) = 1$).

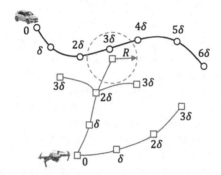

FIGURE 10.5 Illustration of the RRT-based path planning.

10.4.3   One UAV and multiple mobile users

Now, we focus on the case where one UAV periodically surveys multiple mobile users.

The problem is a generalization of some variants of the TSP. Different from the moving-user TSP [19,20], which assumes that the users move at constant speeds in fixed directions, the users in our problem may adjust their headings as well as speeds. Though we can make reasonably accurate predictions on the users' movements, the time needed to intercept a user also depends on how the UAV moves. Thus, the time-dependent TSP [18], which assumes the cost of the time-dependent arcs is known, cannot be used directly to address our problem.

We present a nearest neighbor-based navigation algorithm. Different from the common nearest neighbor approach, which uses distance information as a metric, we consider not only the distance but also the uncertainty level of a user. A user's uncertainty level is modeled as a nondecreasing function of time since its last visit. A typical uncertainty function, denoted by $\gamma(t)$, is shown in Fig. 10.6. Here, the user was last seen at instant t_0. For $t \in [t_0, t_0 + \sigma]$, $\gamma(t) = 0$. After $t_0 + \sigma$, $\gamma(t)$ increases with time. When $\sigma = 0$, $\gamma(t)$ is a monotonically increasing function of time.

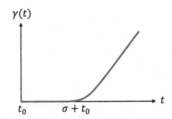

FIGURE 10.6 An illustrative example of the uncertainty level $\gamma(t)$.

We use the symbol $\lambda_j(t)$ to describe how "close" user j is to the UAV. It is defined as follows:

$$\lambda_j(t) := \frac{\gamma_j(t)}{L_j(t)}, \qquad (10.16)$$

where $L_j(t)$ represents the length of the path for the UAV to intercept user j. The path can be found by the method discussed in Section 10.4.2. The metric λ couples the UAV–user distance and the user's uncertainty level. The user with the maximum value of λ is the nearest neighbor.

The navigation method is shown in Algorithm 2. It firstly selects the nearest neighbor. Then, the UAV follows the path to move towards the nearest neighbor. The UAV will choose a new nearest neighbor once the selected nearest neighbor is surveyed. Note that since the users are moving, it is possible that the UAV can have some other users in view before the nearest neighbor. In this case, the UAV will update the users' status but not change the selected nearest neighbor. As a heuristic algorithm, it does not guarantee optimal guidance is obtained. However, since it is a randomized method, it can complete the calculation quickly.

Algorithm 2 Navigating one UAV to survey multiple mobile users.

---
while the time has not reached T **do**
    Apply Algorithm 1 to find the paths to intercept the users.
    Choose the nearest neighbor.
    while the nearest neighbor has not been surveyed **do**
        Keep moving on the planned path.
        Update the users' status if they are surveyed during the movement.
    end while
end while

10.4.4 Multiple UAVs and multiple mobile users

Section 10.4.3 presents how to navigate one UAV to survey multiple mobile users. This subsection extends the navigation algorithm to the scenario having multiple UAVs.

The user revisit time can be further reduced if multiple UAVs are available and their paths are well planned. Then, an important question is how the UAVs coordinate. For coordinating, it requires some extra operations beyond intercepting users. Firstly, the UAVs should exchange their positions and their selected nearest neighbor. Exchanging positions can effectively avoid collision with other UAVs [35,36]. Exchanging the selected nearest neighbor can avoid the case where two UAVs choose the same user as nearest neighbor. Additionally, a UAV should share the status of a user once this user is surveyed. By doing this, all the UAVs have a whole picture of the status of the users, so that they can select their nearest neighbors more effectively. Without this sharing, a UAV may move towards a user which was recently surveyed, and this may increase the average revisit time of the users.

With the shared positions of other UAVs, each UAV can construct a partition of the users. In particular, a user is assigned to UAV i if

$$d_{ij}(t) \leq d_{hj}(t), \forall h = 1, \dots, I, \quad h \neq i, \tag{10.17}$$

where $d_{ij}(t)$ is the horizontal distance between user j and UAV i. Let $S_i(t)$ denote the set of users that are closer to UAV i.

The navigation algorithm for this scenario is summarized in Algorithm 3. Each UAV determines its partition according to the locations of other UAVs and the users. Then, it selects its nearest neighbor in the partition. After that, it starts to pursuit the user. Whenever it views a user other than the nearest neighbor, it shares the status of this user across the team. When the selected nearest neighbor has been surveyed, the UAV repeats the above procedures. The difference between Algorithm 3 and Algorithm 2 is an additional operation in Algorithm 3 to update the partition of a UAV according to the UAVs' and users' locations. Since a UAV needs to calculate the distance from a user to I UAVs, the additional computation complexity is $O(NI)$ for the situation with N users.

Algorithm 3 The navigation algorithm running at a UAV in the team.

while the time has not reached T **do**

    Determine the partition.

    Choose the nearest neighbor in the partition.

    while the nearest neighbor has not been surveyed **do**

        Keep moving on the planned path.

        Update and share the users' status if they are surveyed.

    end while

end while

It is worth pointing out that partitioning the users following (10.17) assigns a user to exactly one group. Then, the nearest neighbor selected by each UAV is unique. Additionally, considering the existence of buildings and the randomness of the RRT-based path planning, two UAVs' paths may be close. In this case, i.e., when two UAVs are within a range of d_{safe}, one of the UAVs, such as the one with the smaller index, continues to fly as planned, while the other UAV tries to avoid the former UAV. To achieve this, the collision avoidance ability should be embedded. Many available UAV products that have such a function, such as DJI Mavic Pro 2, Mavic 2 Zoom, skydio 2, etc., can be used. Fortunately, this is a common ability in most commercial UAVs. When the two UAVs are at least d_{safe} apart, the latter restarts to intercept its nearest neighbor.

Remark 10.4.1. The considered problem in this chapter may not be stable. A simple example is that all the users move away in different directions. Then, the revisit time increases with time.

10.5 Simulations

We show the performance of the proposed navigation algorithms. An urban environment is constructed in MATLAB® with various buildings (Fig. 10.7a). The heights of the buildings are between 30 and 120 m. Five users move on the xy-plane in the environment, and their trajectories are shown in Fig. 10.7b.

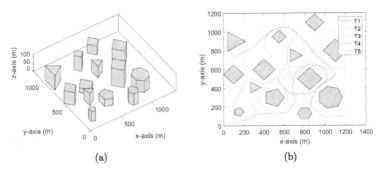

(a)   (b)

FIGURE 10.7 (a) The simulated urban environment. (b) The trajectories of users.

We first consider using one UAV to survey the users. The used parameters are listed as follows: $V^{max} = 10$ m/s, $\Omega^{max} = 0.5$ rad/s, $z = 65$ m, $R = 50$ m, $T = 500$ s, and $\sigma = 0$. It is worth pointing out that the parameters of V^{max} and Ω^{max} depend on the maneuverability of the UAV in use. Our method is not restricted to these parameters. The function $\gamma(t)$ is set to be increasing linearly with time. As will be seen in the results below, since the UAV flight height is higher than some buildings and lower than the others, a UAV needs to avoid the taller buildings but can fly above the lower buildings. To make results more understandable, we assume that during the 500 seconds, the incidence angle does not change. The fixed value can respond to the average value during the 500 seconds. Then, we can precompute the shadow range. Similar to $\mathcal{X}_{building}$, the UAV is not allowed to enter the shadow region. Given an initial state, the trajectory of the UAV is obtained by applying Algorithm 2. In particular, the randomized Algorithm 1 is used to generate UAV trajectories to intercept the five users. In our simulation, for the case with five users, Algorithm 1 (running on a normal computer with an Intel(R) Core(TM) i7-8565U CPU and 8.00G RAM) takes less than 1 second to return five random trajectories. Although the onboard processor may not be as powerful as a normal computer, the algorithm can be coded in the more efficient language C. Moreover, even if the practical computation time may be longer than in the simulation environment, in practice, the UAV can start to compute the random trajectories before it intercepts the intended user. Experimental verification of the proposed methods is left as our future work.

To make the trajectories visible, we demonstrate the 2D views in Fig. 10.8 for each period of 100 seconds. We also record the movements of the UAV and

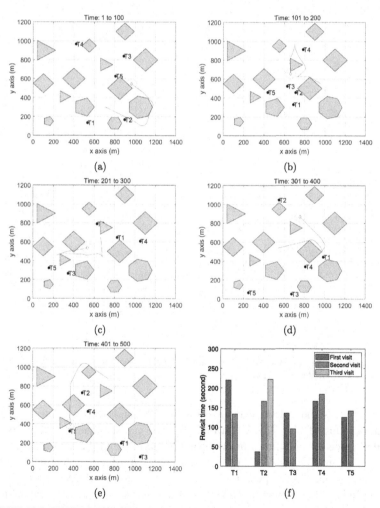

FIGURE 10.8 UAV trajectory when flying at 10 m/s (2D view: https://youtu.be/7cx4jpr0W4I; 3D view: https://youtu.be/eXdSWdH1Yd8).

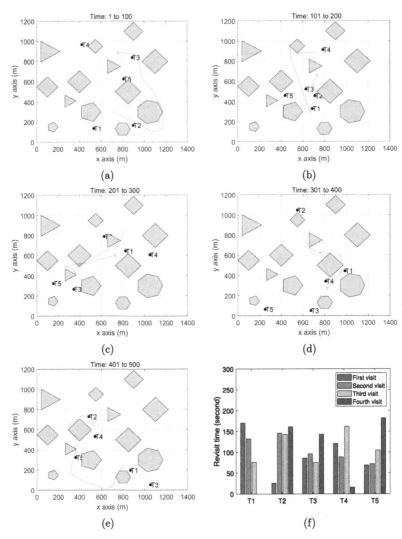

FIGURE 10.9 UAV trajectory when flying at 15 m/s (2D view: https://youtu.be/smvPf-fFdfw; 3D view: https://youtu.be/anSTjbqGoa8).

the users in some videos, and links for both 2D and 3D views are available in the caption of Fig. 10.8. From Fig. 10.8f we can see that the five users are visited twice or three times in the operation period, and the maximum revisit time is about 220 seconds. For the same users' movements, we increase the maximum linear UAV speed to $V^{max} = 15$ m/s, and the UAV's trajectory is shown in Fig. 10.9. From Fig. 10.9f we can see that user 1 was visited three times and all the other users were visited four times in the operation. The maximum revisit time is reduced from 220 seconds to 180 seconds.

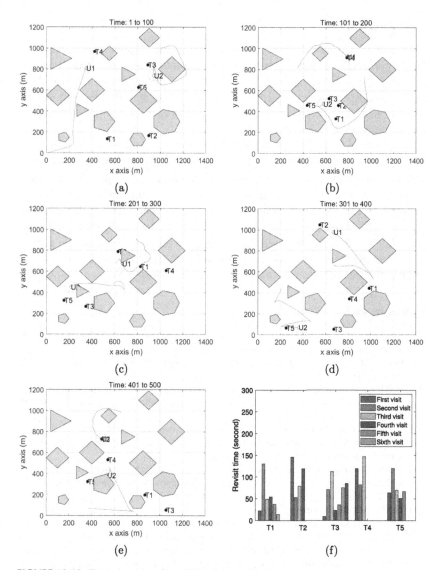

FIGURE 10.10 The trajectories of two UAVs flying at 10 m/s.

Finally, we add one more UAV to survey the same users. Here, $V^{max} = 10$ m/s, and all the other parameters remain the same as above. We show the trajectories of the two UAVs in Fig. 10.10. From Fig. 10.10f we can see that compared to the results using one UAV, the maximum revisit time is reduced from 220 seconds to 150 seconds. The maximum angular speed, i.e., Ω^{max}, has little impact on the maximum revisit time since it only influences the UAV trajectory when the UAV makes a turn.

10.6 Summary

We considered using UAVs to conduct periodical surveillance of moving users. We formulated an optimization problem to minimize the user revisit time. We proposed autonomous navigation algorithms for different cases. Since these algorithms are based on RRT, they inherit the advantage of computational efficiency, and they are promising to be applied in real-time. Simulation results showed their effectiveness.

One limitation of the current work lies in the assumption that the users' trajectories are known. When these trajectories are unavailable, the accuracy of the user position predictions may significantly decrease. Consequently, the UAVs may lose some users. One of our future research directions is to extend the algorithms by including the searching operation. Another direction is to conduct experiments since this is the most effective way to practically verify the computational efficiency of the proposed method. To this end, some mobile ground robots can play the role of users and a small number of UAVs can be tested to conduct the surveillance mission in a laboratory environment.

References

[1] P. Ješke, Š. Klouček, M. Saska, Autonomous compact monitoring of large areas using micro aerial vehicles with limited sensory information and computational resources, in: International Conference on Modelling and Simulation for Autonomous Systems, Springer, 2018, pp. 158–171.

[2] H. Zhang, T. Feng, G.-H. Yang, H. Liang, Distributed cooperative optimal control for multi-agent systems on directed graphs: an inverse optimal approach, IEEE Transactions on Cybernetics 45 (7) (2014) 1315–1326.

[3] Q. Sun, R. Han, H. Zhang, J. Zhou, J.M. Guerrero, A multiagent-based consensus algorithm for distributed coordinated control of distributed generators in the energy internet, IEEE Transactions on Smart Grid 6 (6) (2015) 3006–3019.

[4] J. Sun, Z. Wang, Event-triggered consensus control of high-order multi-agent systems with arbitrary switching topologies via model partitioning approach, Neurocomputing 413 (2020) 14–22.

[5] H. Huang, A.V. Savkin, M. Ding, C. Huang, Mobile robots in wireless sensor networks: a survey on tasks, Computer Networks 148 (2019) 1–19.

[6] H. Huang, A.V. Savkin, Energy-efficient autonomous navigation of solar-powered UAVs for surveillance of mobile ground targets in urban environments, Energies 13 (21) (2020) 5563.

[7] Y. Liu, Q. Wang, H. Hu, Y. He, A novel real-time moving target tracking and path planning system for a quadrotor UAV in unknown unstructured outdoor scenes, IEEE Transactions on Systems, Man, and Cybernetics: Systems 49 (11) (2019) 2362–2372.

[8] S. Wang, F. Jiang, B. Zhang, R. Ma, Q. Hao, Development of UAV-based target tracking and recognition systems, IEEE Transactions on Intelligent Transportation Systems (2019).

[9] R. Wise, R. Rysdyk, UAV coordination for autonomous target tracking, in: AIAA Guidance, Navigation, and Control Conference and Exhibit, 2006, p. 6453.

[10] H. Oh, D. Turchi, S. Kim, A. Tsourdos, L. Pollini, B. White, Coordinated standoff tracking using path shaping for multiple UAVs, IEEE Transactions on Aerospace and Electronic Systems 50 (1) (2014) 348–363.

[11] M. Zhang, H.H. Liu, Cooperative tracking a moving target using multiple fixed-wing UAVs, Journal of Intelligent & Robotic Systems 81 (3–4) (2016) 505–529.

[12] H. Tang, E. Miller-Hooks, Solving a generalized traveling salesperson problem with stochastic customers, Computers & Operations Research 34 (7) (2007) 1963–1987.

[13] P. Toth, D. Vigo, The Vehicle Routing Problem, SIAM, 2002.

[14] A.V. Savkin, H. Huang, Proactive deployment of aerial drones for coverage over very uneven terrains: a version of the 3D art gallery problem, Sensors 19 (6) (2019) 1438.

[15] A.V. Savkin, H. Huang, Asymptotically optimal deployment of drones for surveillance and monitoring, Sensors 19 (9) (2019) 2068.

[16] H. Huang, A.V. Savkin, An algorithm of reactive collision free 3-D deployment of networked unmanned aerial vehicles for surveillance and monitoring, IEEE Transactions on Industrial Informatics 16 (Jan 2020) 132–140.

[17] H. Huang, A.V. Savkin, Reactive 3D deployment of a flying robotic network for surveillance of mobile targets, Computer Networks 161 (2019) 172–182.

[18] J.-F. Cordeau, G. Ghiani, E. Guerriero, Analysis and branch-and-cut algorithm for the time-dependent travelling salesman problem, Transportation Science 48 (1) (2014) 46–58.

[19] C.S. Helvig, G. Robins, A. Zelikovsky, The moving-target traveling salesman problem, Journal of Algorithms 49 (1) (2003) 153–174.

[20] Q. Jiang, R. Sarker, H. Abbass, Tracking moving targets and the non-stationary traveling salesman problem, Complexity International 11 (2005) 171–179.

[21] S. Karaman, M.R. Walter, A. Perez, E. Frazzoli, S. Teller, Anytime motion planning using the RRT, in: 2011 IEEE International Conference on Robotics and Automation, IEEE, 2011, pp. 1478–1483.

[22] Y. Kuwata, J. Teo, G. Fiore, S. Karaman, E. Frazzoli, J.P. How, Real-time motion planning with applications to autonomous urban driving, IEEE Transactions on Control Systems Technology 17 (5) (2009) 1105–1118.

[23] O. Salzman, D. Halperin, Asymptotically near-optimal RRT for fast, high-quality motion planning, IEEE Transactions on Robotics 32 (3) (2016) 473–483.

[24] X. Wang, X. Luo, B. Han, Y. Chen, G. Liang, K. Zheng, Collision-free path planning method for robots based on an improved rapidly-exploring random tree algorithm, Applied Sciences 10 (4) (2020) 1381.

[25] Y. Wu, K.H. Low, An adaptive path replanning method for coordinated operations of drone in dynamic urban environments, IEEE Systems Journal (2020) 1–12.

[26] H. Li, A.V. Savkin, Wireless sensor network based navigation of micro flying robots in the industrial internet of things, IEEE Transactions on Industrial Informatics 14 (8) (2018) 3524–3533.

[27] Y. Kang, J.K. Hedrick, Linear tracking for a fixed-wing UAV using nonlinear model predictive control, IEEE Transactions on Control Systems Technology 17 (5) (2009) 1202–1210.

[28] C. Wang, A.V. Savkin, M. Garratt, A strategy for safe 3D navigation of non-holonomic robots among moving obstacles, Robotica 36 (2) (2018) 275–297.

[29] A.V. Savkin, H. Huang, Optimal aircraft planar navigation in static threat environments, IEEE Transactions on Aerospace and Electronic Systems 53 (Oct 2017) 2413–2426.

[30] A.V. Savkin, H. Huang, W. Ni, Securing UAV communication in the presence of stationary or mobile eavesdroppers via online 3D trajectory planning, IEEE Wireless Communications Letters 9 (8) (2020) 1211–1215.

[31] A.V. Savkin, H. Huang, Bio-inspired bearing only motion camouflage UAV guidance for covert video surveillance of a moving target, IEEE Systems Journal (2020) 1–4.

[32] A.T. Klesh, P.T. Kabamba, Solar-powered aircraft: energy-optimal path planning and perpetual endurance, Journal of Guidance, Control, and Dynamics 32 (4) (2009) 1320–1329.

[33] Y. Zeng, J. Xu, R. Zhang, Energy minimization for wireless communication with rotary-wing UAV, IEEE Transactions on Wireless Communications 18 (4) (2019) 2329–2345.

[34] A.S. Matveev, H. Teimoori, A.V. Savkin, A method for guidance and control of an autonomous vehicle in problems of border patrolling and obstacle avoidance, Automatica 47 (3) (2011) 515–524.

[35] M. Hoy, A.S. Matveev, A.V. Savkin, Algorithms for collision-free navigation of mobile robots in complex cluttered environments: a survey, Robotica 33 (3) (2015) 463–497.

[36] A.V. Savkin, C. Wang, A simple biologically inspired algorithm for collision-free navigation of a unicycle-like robot in dynamic environments with moving obstacles, Robotica 31 (6) (2013) 993–1001.

Index

Printed in the United States
by Baker & Taylor Publisher Services